Molluscan Paleontology of the Chesapeake Miocene

Molluscan Paleontology of the Chesapeake Miocene

Edward J. Petuch • Mardie Drolshagen

CRC Press
Taylor & Francis Group
Boca Raton London New York

CRC Press is an imprint of the
Taylor & Francis Group, an **informa** business

CRC Press
Taylor & Francis Group
6000 Broken Sound Parkway NW, Suite 300
Boca Raton, FL 33487-2742

Printed in the United States of America on acid-free paper
10 9 8 7 6 5 4 3 2 1

International Standard Book Number-13: 978-1-4398-1159-7 (Hardcover)

Library of Congress Cataloging-in-Publication Data

Petuch, Edward J.
Molluscan paleontology of the Chesapeake Miocene / Edward J. Petuch and Mardie Drolshagen.
p. cm.
Includes bibliographical references and index.
ISBN 978-1-4398-1159-7 (alk. paper)
1. Mollusks, Fossil--Chesapeake Bay (Md. and Va.) 2. Paleontology--Miocene. 3. Paleontology--Chesapeake Bay (Md. and Va.) I. Drolshagen, Mardie, 1952- II. Title.

QE801.P439 2010
564--dc22 2009011500

Visit the Taylor & Francis Web site at
http://www.taylorandfrancis.com

and the CRC Press Web site at
http://www.crcpress.com

To Linda, Eric, Brian, and Jennifer Petuch and to the memories of Matthew Barton and Robert Gernant

Contents

Acknowledgments

Without the help and interest of the following individuals, this book would never have been produced. To all of them, we are greatly indebted. For assistance in field collecting, for allowing us access through their properties, and for the donation of valuable research specimens we thank the following: Donald Asher, Dennis Sargert, Norman Riker, Connie Smith, Amy Plummer, Joseph Turner, Robert Schmidt, the Chapman Family, James and Bonnie Vautrot, Robert Shaw, Dr. Geerat Vermeij, Dr. Elizabeth Dudley, Dr. Bretton Kent, Linda Petuch, Edward Mattingly, Beth Permelia, Dahl Daniel, and Eric Kendrew. For reviewing, editing, and helpful comments, we thank Dr. M. G. Harasewych (Smithsonian Institution, Washington, DC), Dr. John Talent (Emeritus, Macquarie University, New South Wales, Australia), Dr. Gregory Dietl (Paleontological Research Institute, Ithaca, New York), and Dr. Anton Oleinik (Florida Atlantic University, Boca Raton, Florida). Special thanks go to Clifford Swearingen for the pen-and-ink drawings on the title page and throughout the book and to Dr. Anton Oleinik for drafting the stratigraphic columns.

Authors

Edward J. Petuch was born in Bethesda, Maryland, in 1949. Being raised in a Navy family, he spent many of his childhood years collecting living and fossil shells in such varied localities as California, Puerto Rico, Chesapeake Bay, and Wisconsin. His early interests in paleontology and marine biology eventually led to B.A. and M.S. degrees in zoology from the University of Wisconsin–Milwaukee. While in Wisconsin, his thesis work concentrated on the molluscan biogeography of coastal West Africa. There, he traveled extensively in Morocco, Western Sahara, Senegal, Gambia, Sierra Leone, and the Cameroons. Continuing his education, Petuch studied marine biogeography under Gilbert Voss at the Rosenstiel School of Marine and Atmospheric Science, University of Miami.

During that time, his dissertation work involved intensive collecting and working on shrimp boats in Colombia, Venezuela, Barbados, the Grenadines, and Brazil. After receiving his Ph.D. in oceanography in 1980, Petuch undertook two years of postdoctoral research on paleoecology with Geerat Vermeij at the University of Maryland. There, he also held a research associateship with the Department of Paleobiology, National Museum of Natural History, Smithsonian Institution, and conducted intensive field work on the Plio–Pleistocene fossil beds of Florida and the Miocene of Maryland.

Petuch has also collected fossil and living mollusks in Australia, Papua–New Guinea, the Fiji Islands, French Polynesia, Japan, the Mediterranean coasts of North Africa and Europe, the Bahamas, Mexico, Belize, Nicaragua, and Uruguay. This research has led to the publication of almost 100 papers. His 10 previous books (*Atlas of Living Olive Shells of the World* (with Dennis Sargent), *New Caribbean Molluscan Faunas, Neogene History of Tropical American Mollusks, Field Guide to the Ecphoras, Edge of the Fossil Sea, Atlas of Florida Fossil Shells, Coastal Paleoceanography of Eastern North America, Cenozoic Seas: The View from Eastern North America, Geology of the Everglades and Adjacent Areas* (with Charles E. Roberts), and *The Geology of the Florida Keys and Everglades: An Illustrated Field Guide to Florida's Hidden Beauty*) are well-known reference texts within the malacological and paleontological communities. Presently, Petuch is a professor of geology in the Department of Geosciences, Florida Atlantic University, Boca Raton, Florida. He resides in Lake Worth, Florida, with his wife Linda and three children, Eric, Brian, and Jennifer. When not collecting or studying mollusks, Petuch leads an active career as a musician and member of the university-affiliated Cuvier Trio, playing recorders and the harpsichord and specializing in Baroque and Renaissance music.

Mardie Drolshagen was born in Detroit, Michigan, in 1952, but soon the family pulled up stakes and moved to sunny southern Florida, where she grew up making daily visits to the ocean. Her father was a WWII photographer and never lost his love for the camera, and he instilled in her the great love of photography that she has today. She started her own photography business, Black Diamond Photography, in 1974, not only taking photos but also teaching photography. Through the years she has worked as a photojournalist for the South Florida Newspaper Network and a photography instructor at Okeeheelee Middle School and Palm Beach Community College Continuing Education.

In 1997 she came to Florida Atlantic University, where she spent the next eleven years as their head photographer. There, she was responsible for aerial, research (technical in-studio and location), architectural, campus life, events, and theater photography.

Her published works include the photography for *The Geology of the Everglades and Adjacent Areas* (2007), *Cenozoic Seas: The View from Eastern North America* (2004), and *The Geology of the Florida Keys and Everglades: An Illustrated Field Guide to Florida's Hidden Beauty* (2008).

Mardie has recently traded in her ocean living for a log cabin in the mountains of western North Carolina.

Introduction

FOSSIL SHELL COLLECTING IN THE MARYLAND AND VIRGINIA MIOCENE

The coastal plain areas of Maryland and Virginia, encompassing Chesapeake Bay and myriad tide water river systems, have long been known to contain some of the richest and best-preserved fossil shell beds found anywhere in the United States. Of these fossil beds, which range from the Paleocene to Pleistocene Epochs, the most prominent and best-developed are those from the Miocene Epoch (5–23 million years BP). The Miocene was a time of wildly fluctuating climates in the Chesapeake Bay area (Petuch, 2004), with some intervals (the early Tortonian Age, 10 million years BP) being warm and subtropical and others (the late Serravallian Age, 12 million years BP) being cold and temperate. The changing environments caused by these fluctuating climates had a tremendous biological impact on the Miocene mollusks. This is reflected in the patterns of rapid evolution and extinction that are beautifully preserved in the fossil record of the Chesapeake Bay area. The amazing array of fossil shells found in exposures of the Chesapeake Miocene makes the area one of the great natural wonders of the eastern United States.

Since coastal Maryland and Virginia are covered by thick deposits of very young Pleistocene and Holocene sands and clays, the older Miocene deposits are only exposed in eroded areas, such as along Chesapeake Bay or along the shores of large rivers. The premier Miocene shell collecting locality is found along the Calvert Cliffs (Figure 1) of Calvert County, Maryland. Here, in a 30-mile stretch of coastline running from Chesapeake Beach in the north to Drum Point near Solomons in the south, three geological formations (the Calvert, Choptank, and St. Mary's Formations) containing 10 richly fossiliferous beds are exposed in the sides of the cliffs (Figure 2). The south shore of the Patuxent River, St. Mary's County, Maryland, also has extensive areas of large cliffs that contain rich layers of beautifully preserved fossil shells. These occur along a 15-mile stretch of Patuxent shoreline running from Golden Beach to the Patuxent River Bridge opposite Solomons. Over 300 species of fossil shells can be collected from the Calvert and Patuxent cliffs alone.

Another important Maryland fossil shell collecting site is found along the St. Mary's River, St. Mary's County. Here, in low cliffs on both sides of the river (Chancellor Point and Windmill Point), an additional 120 species of shells can be collected. The exposures of the St. Mary's Formation found along the St. Mary's River contain the richest and best-preserved molluscan fauna of the entire Chesapeake Miocene (see Chapter 6). Other well-known collecting sites are found near Pope's Creek, Charles County, in the adjacent Potomac River, and along the Choptank River (Talbot and Caroline Counties) of the Maryland eastern shore (Delmarva Peninsula). By collecting in these few areas alone, over 500 species of Miocene mollusks can be amassed, making the Chesapeake Bay area of Maryland one of the most productive fossil localities in the world.

Because the geological formations of Maryland and Virginia dip to the south, the shell beds exposed along the shorelines become progressively younger as one heads down the

FIGURE 1 View of the Calvert Cliffs on Chesapeake Bay, near Chesapeake Beach, Calvert County, Maryland. Here, the Fairhaven Member (Shattuck Zone 3) of the Calvert Formation can be seen at the base of the cliff. Above this dark-colored layer are the lighter-colored sediments of the Plum Point Member of the Calvert Formation (Shattuck Zones 4 through 10). This locality is typical of the northern Calvert Cliffs.

coast. Because of this, the youngest Miocene beds (late Tortonian and Messinian Ages, 6–8 million years BP) make up the majority of the exposures seen in northern Virginia.

Only two areas of Virginia contain extensive beds of Miocene fossils, these being a 15-mile stretch along the south shore of the Rappahannock River, from Bowler's Wharf to Urbanna, Middlesex County (Figure 3) and a 40-mile stretch along the south shore of the James River, from Hopewell to west of Portsmouth, Prince George, Surry, and Isle of Wight Counties. Here, the fossil beds are exposed in low cliffs that line the river edges, and shells that weather out of these layers often litter the river beaches (Figure 4). Other important Virginia Miocene sites include the White Oak Landing area along the Mattaponi River, King William County, and at the Nomini Cliffs near Colonial Beach, Westmoreland County.

In this book, we have assembled all the previous research on the molluscan systematics of the Chesapeake Bay region and produced a synoptic field guide. Considering the richness and accessibility of the Maryland and Virginia Miocene shell beds, it seems remarkable that fewer than two dozen people have ever described new species from these strata over the past 185 years. Of these, only eight workers have ever undertaken faunal surveys that have yielded large numbers of new taxa. These include Say (1824), Conrad (1830–1869), Whitfield (1894; for southernmost New Jersey), Martin (1904), Glenn (1904), Gardner

FIGURE 2 View of the Calvert Cliffs on Chesapeake Bay, 4 miles south of Chesapeake Beach, Calvert County, Maryland. Here, most of the upper Calvert Formation is exposed, including the Plum Point Member (Shattuck Zones 4 through 13; at the bottom and middle of the cliff) and the Calvert Beach Member (Shattuck Zones 14 through 16; at the top of the cliff). The beach can be seen to be filled with fossil mollusks that have weathered out of the cliff face. Rare species, such as *Ecphora mattinglyi* Petuch, 2004 (see Chapter 4) can be collected in blocks of sediments that have fallen off the upper layers.

(1944, 1948), Ward (1985, 1992; for Maryland and Virginia and 1998; for Delaware), and Petuch (1988-2004). Others have been more specialized, describing new species from select molluscan families. Some of these include Green (1830; Conidae), Emmons (1858), Wagner (1839), Dall (1892, 1898), Tucker (1934, 1936; Pectinidae), Schoonover (1941), Mansfield (1929, 1936; Pectinidae), Oleksyshyn (1959, 1960), Gibson (1962; Turridae), Ward and Blackwelder (1975; Pectinidae), Ward and Gilinsky (1988; Muricidae), and Allmon (1990; Buccinidae). The combined research of all these workers has led to the description of over 500 species of mollusks from the Chesapeake Miocene. Future exploration of the cliffs and rivers will doubtlessly lead to the discovery of many more new and interesting species.

Because of recent increased housing development in coastal Virginia and Maryland, many classic fossil collecting localities are now closed to public access. An exception is found at the town of Chesapeake Beach, Calvert County, where a public beach and parking lot allows entry to the northern Calvert Cliffs. Just south of the public beach, it is possible to walk along the base of the cliffs for at least 5 miles. Huge sections of the Calvert Formation are exposed here, and many rare and desirable fossil shells are found on a regular basis.

FIGURE 3 View of the low cliffs along the south side of Urbanna Creek, a branch of the Rappahannock River near Urbanna, Middlesex County, Virginia. Here, the Claremont Manor Member of the Eastover Formation is exposed at the low tide line (covered by dead trees), and the Cobham Bay Member of the Eastover Formation can be seen higher up on the cliff (bands of light-colored shells). Rare species such as *Ecphora whiteoakensis* Ward and Gilinsky, 1988 and *Gradiconus spenceri* Ward, 1992 can be found in these layers.

Another public access site is found farther south at Matoaka Cottages near St. Leonards, Calvert County. Here, collectors have access to over 10 miles of the central Calvert Cliffs (including Governors Run and Calvert Beach) and huge exposures of the upper Calvert Formation and the entire Choptank Formation. Some of the rarest and most beautiful Miocene fossil shells, such as *Chesathais lindae* Petuch, 1988 (Chapter 5, Figure 5.2G,H; cover), have been collected within 2 miles of Matoaka Cottages.

The easiest and safest way to reach the exposures along the Calvert Cliffs, and along the Maryland and Virginia rivers, is with a shallow-draught boat or zodiac. For a nominal fee, local marinas and boat yards will allow you to launch your boat and will provide secure parking facilities. Along the northern Calvert Cliffs, the marina near Breezy Point, Calvert County, is the ideal launching spot for exploring the rich fossil exposures at Plum Point and Dares Beach (with exposures of the Plum Point and Calvert Beach Members of the Calvert Formation). At the southern end of the Calvert Cliffs, at Solomons, Calvert County, the local marina is located just south of the rich Little Cove Point beds of the St. Mary's Formation, at the base of the cliffs at the Chesapeake Ranch Club. The Solomons marina is also the ideal launching spot for exploring the cliffs along the south side of the

FIGURE 4 Close-up of the beach at low tide along Urbanna Creek, near Urbanna, Middlesex County, Virginia. Here, the beach is covered with a pavement of fragments of the pearly oyster *Isognomon* (*Hippochaeta*) *sargenti* Petuch and Drolshagen, new species, and single valves of the frilled oyster *Mansfieldostrea geraldjohnsoni* Ward, 1992. These oysters have weathered out of the dense shell beds of the Cobham Bay Member of the Eastover Formation that are exposed in the adjacent cliffs. Fossil-covered beaches such as this are found all along the Rappahannock River area.

adjacent Patuxent River. Such classic fossil localities as Drumcliff and Sandgates (Queen Tree Landing), St. Mary's County, are only a few miles upriver, and these Patuxent cliffs typically contain richly fossiliferous exposures of the Choptank Formation. The marina and public launching facility at St. Mary's City, St. Mary's County, likewise, will allow access to fossil beds along the St. Mary's River (St. Mary's Formation) such as at Windmill Point, Chancellor Point, and St. Inigoes Creek.

In Virginia, along the Rappahannock River, exposures of beautifully preserved fossils of the Eastover Formation (Claremont Manor and Cobham Bay Members) can be reached by boat from the marina at Bowler's Wharf, Middlesex County. Just 5 miles south of Bowler's Wharf, and near Layton's Landing, densely packed shell beds are exposed along the base of the low cliffs, and many new species have been collected here. Another boat launching site along the Rappahannock River is at the marina and yacht basin at Urbanna, Middlesex County. Urbanna offers two main collecting localities that are easily accessible by boat: the cliffs along the Rappahannock, which contain beautiful exposures of the Cobham Bay Member, and the headwaters of the adjacent Urbanna Creek, which contain shell beds (both Claremont Manor and Cobham Bay) in the low cliffs along the southern shore. The Urbanna Creek cliffs have recently yielded many new species of shells, and they may prove to be one of the most important research localities in the entire Chesapeake Miocene.

Armed with this book, a geological hammer, beach sneakers, and a small boat, any fossil collector should be ready to explore the amazingly rich shell beds of the Chesapeake Bay area. Each locality visited will open up an entire new prehistoric world and will yield an endless array of some of the most beautiful and exotic shells found anywhere on Earth. Collectors today are part of a tradition of paleontological exploration that goes back almost 200 years, to the time of Thomas Jefferson and his friend Thomas Say (the Father of American malacology). As the cliffs of the Chesapeake region gradually erode and expose new sections, this tradition of discovery will be kept alive by future generations of both amateur and professional molluscan paleontologists. The Chesapeake Miocene will always be considered to be one of America's greatest paleontological treasures.

1 Miocene Seas of the Chesapeake Bay Area

The geologic formations of the Chesapeake Bay area contain a detailed record of the oceanographic history of the Miocene Epoch. When the Earth entered periods of global warming during the Miocene, the polar ice caps and high-latitude glaciers melted, raising sea levels and flooding the coastal plain areas. In times of global cooling, the ice caps and glaciers rebuilt, sea level plummeted, and the coastal areas again became dry land. During times of marine flooding (transgressive intervals), thick beds of shells were deposited, and their distribution defines the extent of the oceanic incursion. Sea level lows (regressive intervals) are represented by either missing layers (unconformities) or by layers of terrestrial or unfossiliferous near-shore sediments.

The shell beds and geologic units of coastal Maryland and Virginia were deposited within two oceanic systems: the Salisbury and Albemarle Seas (Petuch, 2004:7–11). Although previously referred to as "embayments" (Gibson, 1983; Ward, 1985, 1992), these systems were far more complex than simple Atlantic coastal bays and represented true seas in the oceanographic sense. Although geographically small, the Salisbury and Albemarle Seas fit all the criteria for designations as true seas (Petuch, 2004:1–6): they were bodies of salt water; they occupied geologically discrete basins and were structurally bound on at least three sides; they contained their own unique configurations of currents and water masses; and they contained their own characteristic endemic organisms and ecosystems. The coastal cliffs of Maryland and Virginia represent cross sections through the sea floors of the Salisbury and Albemarle Seas and contain the oceanographic and ecological history of the entire Miocene Epoch.

As major global warming and cooling events took place throughout the Miocene, the Salisbury and Albemarle basins alternately flooded and dried out as sea levels fluctuated. During warm transgressive intervals, thick, highly fossiliferous sea floors were deposited, and these are the geological formations and members that are exposed in the cliffs of the Chesapeake Bay area. As the ocean retreated during cool regressive intervals, deposition ceased and the exposed sea floors were subject to erosion by rivers and streams, producing unconformities and a gap in the oceanic history of Chesapeake Bay. The separate flooding and depositional episodes of each basin, bounded above and below by unconformities, are referred to as "subseas" (Petuch, 2004). During the Miocene, the basins of the Salisbury and Albemarle Seas, together, contained five separate subseas. The early geologists and paleontologists (Glenn, 1904; Martin, 1904; Shattuck, 1904) recognized these five separate sets of sea floors and described them as distinct formations and faunas. The paleoseas and subseas of the Chesapeake Bay area are described in the following sections.

THE SALISBURY SEA

Named for Salisbury, Wicomico County, Maryland (previously referred to as the "Salisbury Embayment"), the Salisbury Sea existed from the late Oligocene (late Chattian Stage) until the late Miocene (early Tortonian Stage), a period of over 15 million years.

Throughout this time, the Salisbury Sea was separated from the Albemarle Sea of North Carolina by a large structural feature, the Norfolk Peninsula (Petuch, 2004) (the "Norfolk Arch" of Ward, 1992) (Figure 1.1). As an oceanic basin, the Salisbury Sea, from its inception, was doomed to rapid filling by terrestrial sediments, the result of discharge from the large Millville Delta along its northern edge (Petuch, 2004) (Figure 1.1). This feature, produced by the proto-Susquehanna or proto-Delaware Rivers, progressively filled the Salisbury Sea with clay, mud, and sand until Tortonian time, when the paleosea was a tiny sliver of its former self.

Within the Salisbury Sea depositional sequence, four subseas can be recognized. The oldest, the Old Church Subsea (named for the Old Church Formation; Petuch, 2004), existed during the late Chattian Oligocene and early Aquitanian Miocene. Little is known about this subsea (Ward, 1985), because only a few small natural outcrops occur and its fossils

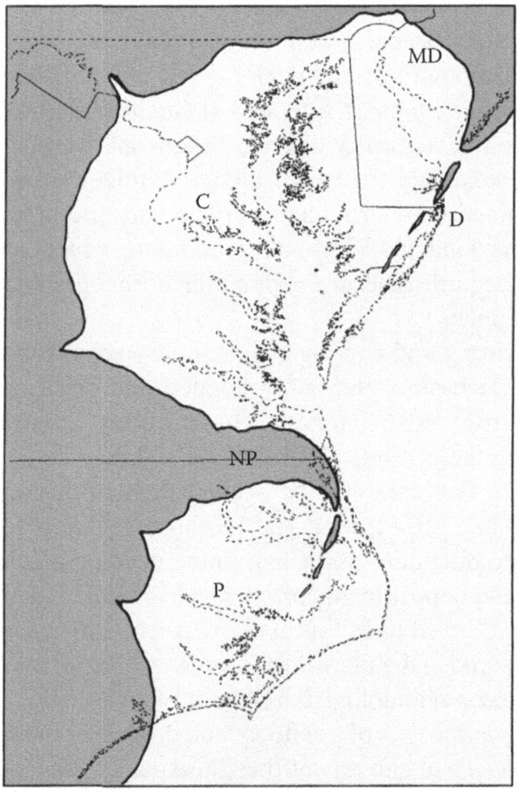

FIGURE 1.1 Approximate configurations of the Calvert Subsea of the Salisbury Sea (C) and the Pamlico Subsea of the Albemarle Sea (P), during the late Burdigalian and Langhian Miocene, superimposed upon the outline of Recent Chesapeake Bay and northern North Carolina. Prominent geographic features include the Millville Delta (MD), the Delmarva Archipelago (D), and the Norfolk Peninsula (NP).

are poorly preserved. For these reasons, the Old Church Subsea and Old Church Formation will not be discussed in this book.

The second subsea, the Calvert Subsea (named for the Calvert Formation; Petuch, 2004) (Figure 1.1), existed during the late Burdigalian and Langhian Ages of the Miocene. Spatially, this was the largest of the four subseas, encompassing an area that extended from Richmond, Virginia in the west, southern New Jersey and Delaware in the north, Norfolk, Virginia in the south, and covering the entire Chesapeake Bay area. At the end of Langhian time, sea levels dropped and the entire Salisbury Sea basin dried out for tens of thousands of years. During this subaerial time, the Millville Delta deposited huge amounts of sediments, filling the northern half of the basin.

The third subsea, the Patuxent Subsea (named for the Patuxent River; Petuch, 2004) (Figure 1.6), was only about one third the size of the previous Calvert Subsea and existed during the middle part of the Serravallian Age. At this time, a mid-Miocene worldwide cooling event was taking place, and the lowered sea levels caused the southern third of the basin to become emergent. Before the beginning of late Serravallian time, sea level dropped precipitously and the Salisbury Sea again was emergent for tens of thousands of years. During this time, the Millville Delta buried much of the Chesapeake Bay area.

At the end of the Serravallian Age, the marine climate began to warm and sea level rose again, flooding the last remnant of the Salisbury Sea basin. This fourth and final Salisbury subsea, the St. Mary's Subsea (named for the St. Mary's River; Petuch, 2004) (Figure 1.10), was a mere remnant of the older subseas, being only about one quarter the size of the Calvert Subsea. The Millville Delta rapidly prograded into the St. Mary's Subsea and completely filled the Salisbury basin by mid-Tortonian time. At the same time that this infilling was taking place, the area of the Norfolk Peninsula was rapidly eroding, due to rising sea levels during the mid-Tortonian warm period. This allowed the Albemarle Sea to connect with the southernmost remnants of the Salisbury Sea, producing a new late Miocene world, the Rappahannock Subsea (Figure 1.11; discussed later in this chapter).

THE CALVERT SUBSEA AND ITS ENVIRONMENTS

During peak Calvert time (Langhian Age), the Salisbury Sea was flooded with warm-temperate water and housed an amazing array of endemic marine ecosystems. Some of the more important and interesting Calvert mollusks, residents of these ecosystems, are shown in Chapter 4. At the beginning of the late Burdigalian, sea levels were still low and water temperatures were not as warm as they were in the Langhian. These stressed ecosystems of the early Calvert Subsea (preserved in the Fairhaven Member of the Calvert Formation; see Chapter 2) contained low-diversity faunas, dominated by the estuarine clam *Caryocorbula elevata* (Conrad, 1838) and rich deposits of diatoms.

During latest Burdigalian time, marine climates began to warm and sea level rose appreciably. The now-drowned Millville Delta system (in present-day southern New Jersey and Delaware; see Ward, 1998) produced a series of shallow, muddy environments that housed their own endemic marine communities. Principal among these was the *Bulliopsis variabilis* Community (Petuch, 2004), which contained many species of mollusks that have not been found anywhere else in the Chesapeake Miocene. Farther south, in what is now northern Chesapeake Bay, immense beds of the gryphaeid oyster *Hyotissa percrassa* (Conrad, 1841) (Figure 1.2) grew in shallow muddy areas adjacent to the delta and supported an interesting fauna of drilling carnivorous muricid gastropods such as *Trisecphora*

FIGURE 1.2 Detail of a bed of the gryphaeid oyster *Hyotissa percrassa* (Conrad, 1841) in Shattuck Zone 4, Plum Point Member of the Calvert Formation, 3 miles south of Chesapeake Beach, Calvert County, Maryland. The ecphora *Trisecphora chamnessi* (Petuch, 1989) has been collected in these oyster bioherms.

chamnessi (Petuch, 1989) (see Chapter 4) (the *Hyotissa percrassa* Community; Petuch, 2004). Shallow muddy lagoonal areas, under deltaic influence, housed immense monoculture beds of the small estuarine bivalve *Caryocorbula elevata* (Figure 1.3), demonstrating the stressed conditions of the brackish water environment associated with the Millville Delta.

As conditions continued to warm and sea level rose, the area of central Chesapeake Bay became more lagoonal in nature, with open oceanic water conditions and normal salinity. Two main types of soft sediment, open-bottom communities evolved here and included one centered on extensive beds of turritellid gastropods (the *Torcula exaltata* Community; Petuch, 2004) and one centered on beds of scallops and other large bivalves (the *Chesapecten coccymelus* Community) (Figure 1.4). These two communities interfingered across the deeper water areas of the early Langhian Calvert Subsea and contained one of the highest levels of biodiversity and species-richness seen in the Chesapeake Miocene (see Chapter 4).

Toward the end of Langhian time, the marine climate again began to deteriorate, and lower sea levels and colder water conditions prevailed. Many of the tropical-subtropical genera found in the *Torcula exaltata* and *Chesapecten coccymelus* Communities died out at this time, with some never to be found again at such a high latitude (see Chapter 4). On the

FIGURE 1.3 Detail of a bed of the small corbulid bivalve *Caryocorbula elevata* (Conrad, 1838) in Shattuck Zone 6, Plum Point Member of the Calvert Formation, 3 miles south of Chesapeake Beach, Calvert County, Maryland.

shallow, muddy-bottom areas of the late Langhian Calvert Subsea, large beds of the venerid bivalves *Melosia staminea* (Conrad, 1839) and *Mercenaria blakei* Ward, 1992 covered extensive areas and supported an interesting fauna of carnivorous drilling murex shells, primarily the wide-ribbed *Ecphora calvertensis* Petuch, 1988 and *Trisecphora martini* (Petuch, 1988) (the *Ecphora calvertensis* Community; Petuch, 2004). During the very latest Langhian Age, only hardy, impoverished ecosystems survived within the Calvert Subsea. Some of these were dominated by dense beds of the heart-shaped bivalve *Glossus mazlea* subspecies (Figure 1.5) while others contained beds of small mactrid and corbulid bivalves and their primary predators *Ecphora chesapeakensis* Petuch, 1992 and *Ecphora (Planecphora) turneri* Petuch, 1992 (the *Ecphora chesapeakensis* Community; Petuch, 2004).

THE PATUXENT SUBSEA AND ITS ENVIRONMENTS

The cool marine climate at the end of Langhian time continued to intensify in the following Serravallian Age, culminating in the "Mid-Miocene Cold Time" and the resulting "Transmarian Extinction" (Petuch, 1993; 2004). As sea level continued to drop in the early Serravallian, the Millville Delta was reactivated and filled the northern half of the Salisbury Sea with a huge deltaic and estuarine system. By the mid-Serravallian, the resultant new

FIGURE 1.4 Detail of a molluscan assemblage from the *Chesapecten coccymelus* Community in Shattuck Zone 10, Plum Point Member of the Calvert Formation. At this locality 3 miles south of Chesapeake Beach, Calvert County, Maryland, the yellowish sand can be seen to be packed with broken fragments of the scallop *Chesapecten coccymelus* (Dall, 1898) and the glossid clam *Glossus mazlea* (Glenn, 1904). Two specimens of the astartid clam *Astarte cuneiformis* Conrad, 1840 can be seen in the center of the photograph.

subsea, the Patuxent Subsea (Figure 1.6), was less than half the size of the Calvert Subsea. The Delmarva Archipelago of islands (named for the Delmarva Peninsula; Petuch, 2004) was also well developed at this time, effectively isolating the Salisbury Sea from the open Atlantic. Throughout the Serravallian, the Patuxent Subsea was essentially a large, shallow, mud- and sand-bottomed lagoon (Figure 1.7).

Because of the limited types of substrates and depth ranges, the Patuxent Subsea housed only a few types of marine ecosystems. In the early Serravallian, the mud and sand bottom areas of the central lagoon contained huge beds of large bivalves, primarily the lucinid *Lucinoma contracta* (Say, 1824) and the venerid *Mercenaria cuneata* (Conrad, 1867), and these supported a large fauna of carnivorous drilling muricid gastropods such as *Trisecphora smithae* (Petuch, 1988), *Ecphora sandgatesensis* Petuch, 1989, and *Ecphora (Planecphora) vokesi* Petuch, 1988 (see Chapter 5) (the *Trisecphora smithae* Community; Petuch, 2004). Small scattered beds of the scallop *Chesapecten nephrens* subspecies and large bioherms of the ahermatypic coral *Astrhelia palmata* (Goldfuss, 1826) and the pearly oyster *Isognomon (Hippochaeta) torta* (Say, 1820) also occurred on the shallow muddy bottoms.

During the mid-Serravallian, two main types of marine communities dominated the mud and sand sea floors of the Patuxent Subsea. One was composed of extensive beds of

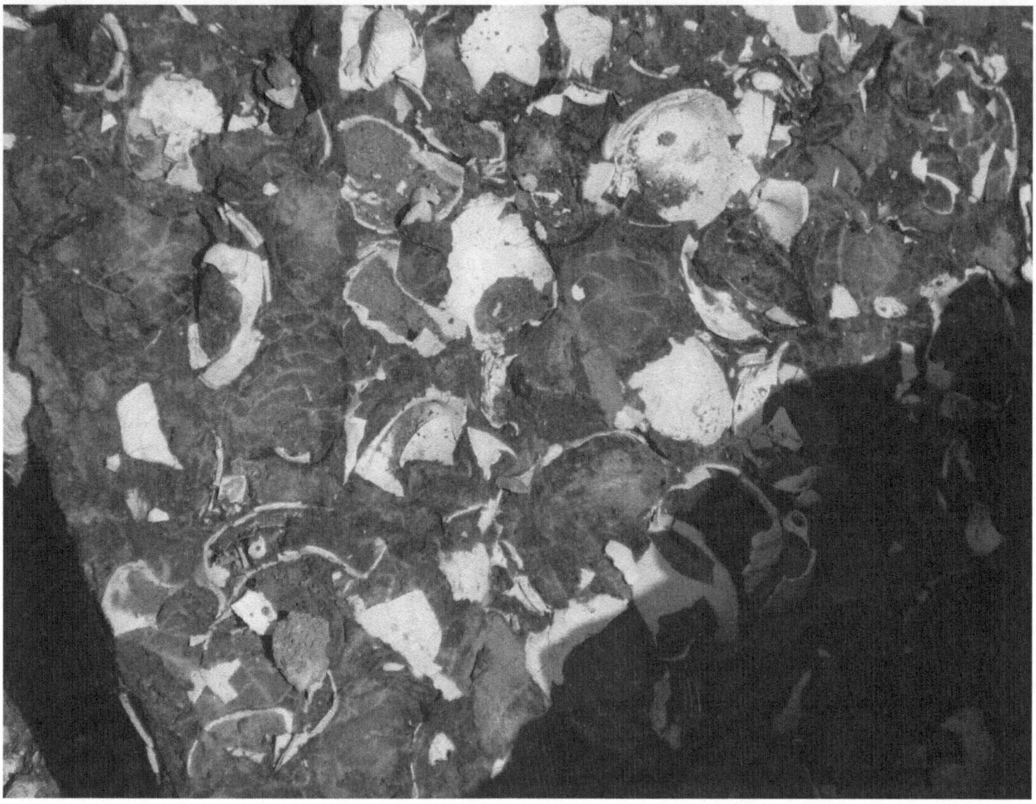

FIGURE 1.5 Detail of a bed of the glossid clam *Glossus mazlea* (Glenn, 1904) in Shattuck Zone 11, Plum Point Member of the Calvert Formation, 3 miles south of Chesapeake Beach, Calvert County, Maryland.

the large scallop *Chesapecten nephrens* Ward and Blackwelder, 1975 (Figure 1.8) and supported a distinctive fauna of large molluscivorous gastropods such as *Busycotypus choptankensis* (Petuch, 1993), *Sycopsis lindae* Petuch, 1988, and *Ecphora rikeri* Petuch, 1988 (the *Chesapecten nephrens* Community; Petuch, 2004) (see Chapter 5). The other, in muddier areas, was dominated by dense beds of several types of smaller bivalves, including *Marvacrassatella turgidula* (Conrad, 1843), *Glossus marylandica* (Schoonover, 1941), and *Bicorbula drumcliffensis* (Oleksyshyn, 1960) (the *Bicorbula drumcliffensis* Community; Petuch, 2004). Feeding on this rich bivalve fauna was a host of carnivorous drilling muricid gastropods, including *Ecphora meganae* Ward and Gilinsky, 1988 (Figure 1.9), *Ecphora (Planecphora) choptankensis* Petuch, 1988, and *Chesathais lindae* Petuch, 1988 (see Chapter 5).

At the end of the Serravallian Age, the marine climate became colder and sea levels began to drop. This stressed environment supported only an impoverished molluscan fauna that was far less species rich than those of the mid-Serravallian. Dominating the open, muddy bottoms were huge beds of the turritellid gastropod *Mariacolpus octonaria* (Conrad, 1863) and the bivalves *Marvacrassatella marylandica* (Conrad, 1832), *Chesacardium patuxentia* (Glenn, 1904), and *Chesapecten monicae* Ward, 1992 (the *Mariacolpus octonaria* Community; Petuch, 2004) (see Chapter 5). Feeding on the turritellids and bivalves was a large fauna of carnivorous gastropods including the drilling muricids *Ecphora williamsi*

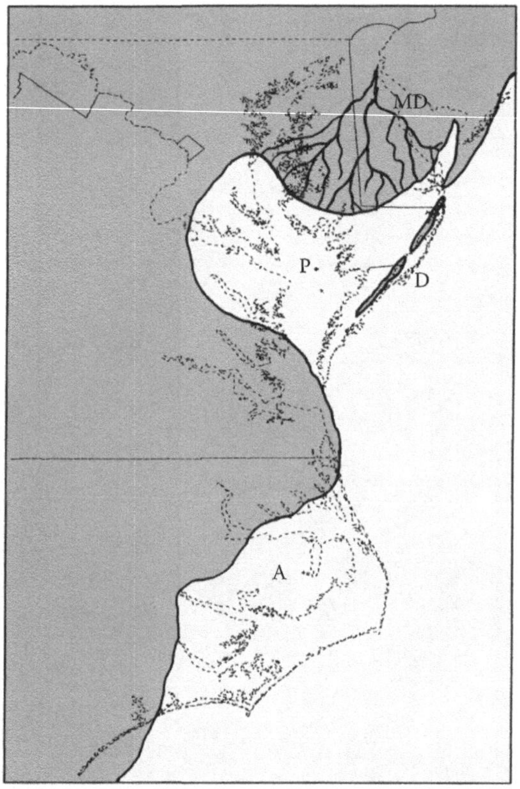

FIGURE 1.6 Approximate configurations of the Patuxent Subsea of the Salisbury Sea (P) and the remnants of the Albemarle Sea (A), during the Serravallian Age of the Miocene, superimposed upon the outline of Recent Chesapeake Bay and northern North Carolina. Prominent geographical features include the Millville Delta (MD) and the Delmarva Archipelago (D). At this time, the Albemarle Sea was filled with sediments and ceased to be a separate oceanographic entity.

Ward and Gilinsky, 1988 and *Ecphora (Planecphora) delicata* Petuch, 1989 and the large busycon whelk *Busycotypus choptankensis* (Petuch, 1993). By the end of the Serravallian, sea level dropped dramatically and the entire Salisbury Sea basin was exposed to subaerial conditions for over 500,000 years.

THE ST. MARY'S SUBSEA AND ITS ENVIRONMENTS

The fourth and final subsea of the Salisbury Sea, the St. Mary's Subsea (Figure 1.10), formed during the early Tortonian warm time. During the late Serravallian sea level drop, when water temperatures were still relatively cool and sea level was just beginning to rise, the early St. Mary's Subsea contained only shallow, muddy-bottomed areas with low diversity ecosystems. These were dominated by beds of small scallops (*Chesapecten* species) and turritellid gastropods (*Mariacolpus* species), and supported several large drilling predators, primarily the moon snail *Euspira* and the muricid *Ecphora* (the *Ecphora conoyensis* and *Ecphora asheri* Communities; Petuch, 2004). As water temperatures rose in the early Tortonian, the turritellid- and pectinid-based communities became more extensive and increased in species diversity (the *Mariacolpus covepointensis* and *Chesapecten covepointensis* Communities; Petuch, 2004).

Later in the early Tortonian, water temperatures reached their maxima and rich subtropical ecosystems flourished on the open muddy sea floors south of the Millville Delta. In deeper water areas, extensive beds of large *Chesapecten* scallops supported a wide array of large gastropod predators, including busycon whelks (*Coronafulgur, Busycotypus, Sycopsis,* and *Turrifulgur*), volutes (*Volutifusus*), and horse conchs (*Conradconfusus, Mariafusus,* and *Pseudaptyxis*) (the *Chesapecten santamaria* Community; Petuch, 2004). This scallop-based community interfingered with dense beds of the turritellid gastropod *Mariacolpus lindajoyceae* Petuch, 2004 and its associated predators (the *Mariacolpus lindajoyceae* Community; Petuch, 2004). In shallower water areas, on muddy sand bottoms, beds of large and small venerid bivalves dominated the marine communities (the *Mercenaria tetrica* Community and the *Lirophora alveata* Community; Petuch, 2004). All four of these molluscan-based communities contained southern tropical genera that had migrated northward into the St. Mary's Subsea during the early Tortonian warm time. Some of these, which had never been found in Maryland before or after that time, include the cone shell *Gradiconus,* the miter shell *Mitra,* the tun shell *Eudolium,* and the marginellid *Dentimargo* (see Chapter 6).

One of the more interesting marine environments of the St. Mary's Subsea was represented by the intertidal mud flats found along the southern edge of the Millville Delta. Here, small nassariid mud snails (*Ilyanassa*) and the small Transmarian Province endemic whelk genus *Bulliopsis* dominated the molluscan ecosystem (the *Bulliopsis quadrata* Community; Petuch, 2004). These small scavengers fed on detritus carried down to the St. Mary's Subsea by deltaic levees and rivers. These estuarine swamps and mud flats were also the home of the giant razor-toothed crocodilian *Thecachampsa sicaria* Cope, 1869.

At the end of the early Tortonian, sea levels again began to drop and the climate cooled. For a period of at least 200,000 years, the ocean regressed and the St. Mary's Subsea dried out. This was reversed for a short time at the beginning of the middle Tortonian, when the St. Mary's Subsea was again briefly flooded. This final marine incursion, preserved in the Chancellor Point Sandstone (see Chapter 6), contained only a single low-diversity molluscan-based ecosystem (the *Gradiconus asheri* Community; see Petuch, 2004). Within this community, the characteristic vermivorous cone shell *Gradiconus asheri* occurred along with beds of the ark shell *Dallarca chancellorensis* Petuch and Drolshagen, new species (see Systematic Appendix), the scallop *Chesapecten* species, and the small venerid bivalve *Lirophora alveata* subspecies (see Chapter 6). Within a few tens of thousands of years, sea level again regressed and this last St. Mary's fauna became extinct. This was the final marine ecosystem of the Salisbury Sea.

THE ALBEMARLE SEA

Named for the Albemarle Sound of North Carolina, the Albemarle Sea existed from the Oligocene to the early Pleistocene, a greater time span than that covered by the Salisbury Sea. During the Burdigalian and Langhian Ages, the Albemarle Sea was separated from the Salisbury Sea by the wide Norfolk Peninsula structural arch (Petuch, 2004; Ward, 1985, 1992). Although separated geographically, the two seas shared essentially the same molluscan fauna and same types of environments. This is readily demonstrated by the large number of species in common between the faunas of the Calvert Formation (Calvert Subsea, Salisbury Sea) and the Pungo River Formation of North Carolina (Pamlico Subsea,

FIGURE 1.8 Detail of an exposure of Shattuck Zone 18, Drumcliff Member of the Choptank Formation, north of Matoaka Cottages, St. Leonard, Calvert County, Maryland. Here, the large scallop *Chesapecten nephrens* Ward and Blackwelder, 1975 can be seen exposed in the side of the cliff, along with fragments of large bivalves such as *Mercenaria cuneata* subspecies and *Chesacardium blackwelderi* Ward, 1992. This assemblage is typical of the *Chesapecten nephrens* Community.

Albemarle Sea) (Petuch, 2004:95, 96–99). By late Serravallian and early Tortonian time, however, the Norfolk Peninsula had eroded into a broad, flat coastal plain and had filled in the Pamlico Subsea basin (Figure 1.10). For almost 7 million years, the Albemarle Sea ceased to exist and was replaced by terrestrial conditions.

After the large regressive interval during the Serravallian-to-mid-Tortonian, sea level rose rapidly and quickly eroded away the flat Virginia coastal plain and the remnants of the Norfolk Peninsula. This reexcavated the Albemarle basin and formed a connection with the last remnants of the St. Mary's Subsea to produce a new oceanic system, the Rappahannock Subsea of the Albemarle Sea (Figure 1.11). Many of the surviving St. Mary's mollusks migrated into the Rappahannock Subsea and evolved into new species and species

FIGURE 1.7 **(Facing page)** View of the Calvert Cliffs south of Governor Run, Calvert County, Maryland, showing the rich fossil shell beds of Shattuck Zone 18, Drumcliff Member of the Choptank Formation. Here, such rare and interesting species as the ecphoras *Chesathais lindae* Petuch, 1988 and *Ecphora rikeri* Petuch, 1988, the busycon whelk *Sycopsis lindae* Petuch, 1988, and the cancellariid *Ventrilia biplicifera* (Conrad, 1841) are occasionally collected.

FIGURE 1.9 Detail of an exposure of Shattuck Zone 18, Drumcliff Member of the Choptank Formation, north of Matoaka Cottages, St. Leonard, Calvert County, Maryland. Here, a broken specimen of the ecphora muricid *Ecphora meganae* Ward and Gilinsky, 1988 lies in contact with the large crassatellid bivalve *Marvacrassatella turgidula* (Conrad, 1843). Both specimens are surrounded with broken fragments of *Mercenaria* clams and the turritellid *Mariacolpus octonaria* subspecies.

complexes. Because of the similarities between the shells of the Eastover Formation of Virginia and northern North Carolina (Rappahannock Subsea) and those of the St. Mary's Formation (St. Mary's Subsea), the geological units of the Rappahannock Subsea had been referred to previously as the "Virginia St. Mary's."

THE RAPPAHANNOCK SUBSEA AND ITS ENVIRONMENTS

Named for the Rappahannock River of Virginia (Petuch, 2004), the Rappahannock Subsea existed only from the late Tortonian Age to the early Messinian Age of the Miocene. During the late Tortonian, cool water temperatures prevailed within the subsea and the molluscan fauna was correspondingly impoverished. By the early Messinian, however, warm water conditions dominated and several subtropical molluscan genera migrated into the Albmarle basin from the south (see Chapter 7). These warm water conditions abruptly ceased, and sea level dropped precipitously, in late Messinian time. This abrupt major regression and severe cold time, often referred to as the "Messinian Event," occurred worldwide and may have been triggered by a large meteor impact at Kara-Kul, Tajikistan. After the "Messinian Event," the entire Albemarle Sea basin dried out for over 4 million years.

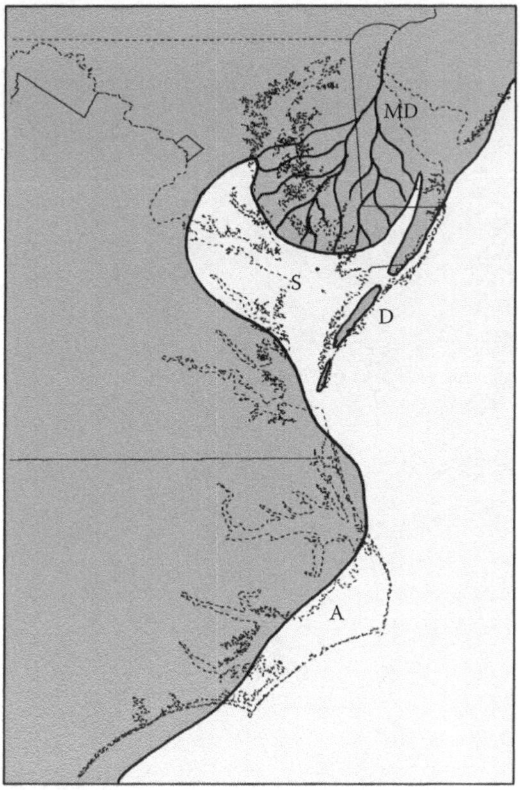

FIGURE 1.10 Approximate configuration of the St. Mary's Subsea of the Salisbury Sea (S) during the early Tortonian Miocene, superimposed upon the outline of Recent Chesapeake Bay. Prominent geographical features include the Millville Delta (MV) and Delmarva Archipelago (D). At this time, the Albemarle Sea of the Burdigalian, Langhian, and early Serravallian is now gone, replaced by a shallow coastal bank, the Albemarle Bank (A).

The northern half of the Rappahannock Subsea was the most geomorphologically complex and consisted of shallow-water areas with mud and sand bottoms. Two chains of low sandy islands separated the subsea into discrete basins (Figure 1.11). One of these island chains, the Delmarva Archipelago, extended southward from the incipient Delmarva Peninsula and separated and protected the main Rappahannock Subsea basin from the open Atlantic. The other island chain, the Bayport Archipelago (Petuch, 2004; referred to as the "Bayport High" by Ward and Blackwelder, 1980), separated the coastal lagoons from the main Rappahannock basin. Behind the Bayport Archipelago, mud and clay sea floors prevailed, while outside the island chain, sand and shell rubble sea floors predominated (Figure 1.12). In the extreme south of the Rappahannock Subsea, a large shallow bank occurred (the Aurora Bank; Petuch, 2004) (Figure 1.11) and contained a much higher percentage of carbonates. Here, and in contemporaneous areas in western Florida (Petuch and Roberts, 2007), the large gryphaeid oyster *Gigantostrea leeana* (Wilson, 1987) formed extensive biohermal structures.

During the late Tortonian cool time, several types of marine communities existed within the Rappahannock Subsea. In the muddy-bottom, shallow lagoon areas behind the Bayport Archipelago, dense beds of the marsh clam *Mulinia rappahannockensis* (Gardner, 1944), the venerid *Mercenaria druidi gardnerae* new subspecies (see Systematic Appendix),

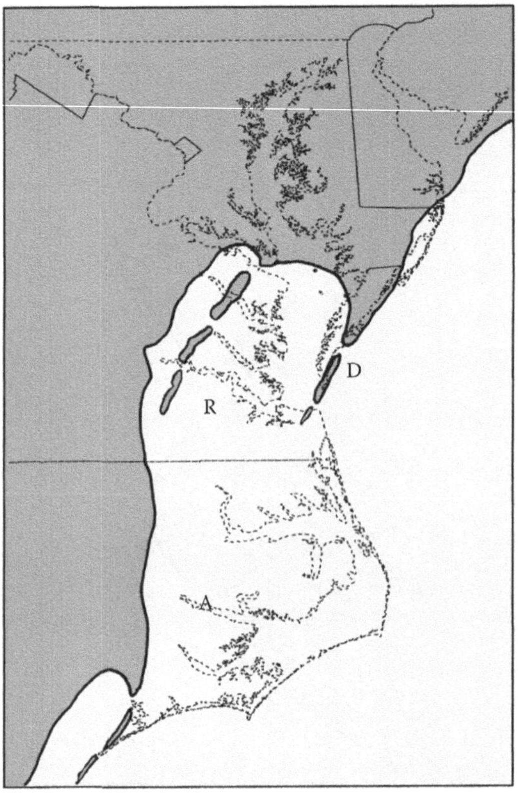

FIGURE 1.11 Approximate configuration of the Rappahannaock Subsea of the Albemarle Sea (R) during the late Tortonian and Messinian Miocene, superimposed upon the outline of Recent Chesapeake Bay and northern North Carolina. Prominent geographical features include the Delmarva Archipelago (D) and the Aurora Bank (A). The Bayport Archipelago of islands can be seen to extend along the northwestern shore.

and the arcid *Dallarca rotunda* Ward, 1992 accumulated. These were the principal prey items of the carnivorous gastropods *Ecphora whiteoakensis* Ward and Gilinsky, 1988 and the busycon whelk *Coronafulgur kendrewi* Petuch, 2004 (the *Coronafulgur kendrewi* Community; Petuch, 2004) (see Chapter 7). In deeper-water areas, in the open sound between the Bayport and Delmarva Archipelagos, the sand sea floor was dominated by beds of the scallop *Chesapecten ceccae* Ward, 1992, the turritellid gastropod *Mariacolpus carinatus* (Gardner, 1948), and the bittersweet clam *Glycymeris virginiae* (Dall, 1898) (the *Chesapecten ceccae* Community; Petuch, 2004).

The warm time that ushered in the latest Tortonian and early Messinian saw an increase in the species richness and diversity of the Rappahannock Subsea ecosystems. In the muddy coastal lagoon behind the Bayport Archipelago, and also on intertidal sand banks throughout the open sound area, the small marsh clam *Mulinia rappahannockensis* (Gardner, 1944) again formed thick beds (Figure 1.13). Unlike the *Mulinia* beds of the late Tortonian, the Messinian beds contained a much richer and more diverse molluscan community, typically containing the large cockle shell *Chesacardium blountense* (Mansfield, 1932), the arcid *Dallarca carolinensis* (Dall, 1898), and the surf clam *Spisula bowlerensis* Gardner, 1943. These were the principal prey items for a large number of carnivorous gastropods, some of which included the muricids *Ecphora kochi* Ward and Gilinsky, 1988 and *Urosalpinx rappahannockensis*

FIGURE 1.12 Detail of an exposure of the Eastover Formation along the eastern side of Urbanna Creek, 4 miles from Urbanna, Middlesex County, Virginia. The lower Claremont Manor Member can be seen exposed along the tide line, and the upper Cobham Bay Member is exposed higher up on the cliff face. The large shells exposed in the Cobham Bay Member are mostly the giant pearly oyster *Isognomon (Hippochaeta) sargenti* Petuch and Drolshagen, new species.

(Gardner, 1948) and the busycon whelk *Coronafulgur kendrewi* Petuch, 2004 (all part of the *Mulinia rappahannockensis* Community; Petuch, 2004). As in the muddy deltaic areas of the St. Mary's Subsea, *Bulliopsis* snails (*Bulliopsis bowlerensis* Allmon, 1990) were also abundant on the intertidal mud flats behind the Bayport Archipelago.

On clean sand bottoms in deeper-water areas of the open sound, an extremely rich molluscan community evolved during Messinian time. This ecosystem, the *Oliva idonea* Community (Petuch, 2004), contained large beds of the giant pearly oyster *Isognomon (Hippochaeta) sargenti* Petuch and Drolshagen, new species (see Systematic Appendix) (Figure 1.14) and the bivalves *Astarte rappahannockensis* Gardner, 1944, *Costaglycymeris mixoni* Ward, 1992 (Figure 1.15), and *Striarca centenaria* (Say, 1824) (see Chapter 7). Living on open sand patches between the bivalve beds were two characteristic tropically derived gastropods, the olivid *Oliva (Strephona) idonea* Conrad, 1839 (namesake of the community) and the cone shell *Gradiconus spenceri* (Ward, 1992). Interfingering with the bivalve, olive, and cone shell community were immense beds of giant scallops, including *Chesapecten middlesexensis* (Mansfield, 1936) (Figure 1.16), *Carolinapecten urbannaensis* (Mansfield, 1929), and *Placopecten principoides* (Emmons, 1858), and their associated organisms (the *Placopecten principoides* Community; Petuch, 2004).

FIGURE 1.13 Detail of a bed of the marsh clam *Mulinia rappahannockensis* (Gardner, 1944) exposed in a cliff on Urbanna Creek, 4 miles from Urbanna, Middlesex County, Virginia. This assemblage is typical of the *Mulinia rappahannockensis* Community of the Cobham Bay Member of the Eastover Formation.

MIOCENE MARINE BIOGEOGRAPHY

Marine biogeography, the study of the distribution of marine organisms in time and space, is crucial to our understanding of evolution and extinction in the world's oceans. Marine biogeographers such as Valentine (1973), Briggs (1974), and Petuch (2004) have divided the coastal areas of the world into discrete geographical units called "provinces." Controlled primarily by water temperatures, provinces are defined by the "50% Rule." This states that two areas belong to separate molluscan provinces if, upon comparison, at least 50% of their species are different. Molluscan provinces can be further subdivided into subprovinces, using the "30% Rule" (Petuch, 1988; 2004). This states that, if 30% of the molluscan taxa differ between two areas within a province, they can be considered separate subprovinces of that province.

During the Miocene, from the Aquitanian to the early Messinian Ages, two molluscan provinces extended along the eastern coast of the United States. One of these, the **Baitoan Province** (named for the Baitoa Formation of the Dominican Republic, which contains a typical fauna; see Petuch, 2004:29), ranged from northern South America, along Central America, across the Gulf of Mexico and Florida, and northward to the Cape Fear region of North Carolina. The other, the **Transmarian Province** (named for the St. Mary's River of Maryland; Petuch, 1988, 1993, 2004), ranged from Cape Fear northward to the Gulf of Maine.

FIGURE 1.14 Detail of a bed of the giant pearly oyster *Isognomon (Hippochaeta) sargenti* Petuch and Drolshagen, new species exposed in a cliff on Urbanna Creek, 4 miles from Urbanna, Middlesex County, Virginia.

The Miocene Cape Fear area was a large peninsula that jutted out into the Atlantic and was the divergence point for two Miocene oceanic currents. Flowing southward from the north, and giving the Transmarian Province a temperate marine climate, was the cool **Raritan Current** (see Petuch, 1997). Flowing northward from the south was the warm **Arawak Current**, which flooded the Baitoan Province with tropical and subtropical conditions. The two currents met at the Cape Fear Peninsula and flowed eastward into the open Atlantic.

Faunistically, the Transmarian and Baitoan Provinces differed greatly, reflecting the contrasting influences of the Raritan and Arawak Currents. The cooler-water Transmarian Province contained many characteristic endemic gastropod genera (found only within the boundaries of the province), some of which are listed here by family (see Chapters 4, 5, 6, and 7):

Turritellidae
- *Calvertitella* Petuch, 1988 (*C. indenta* species complex)
- *Mariacolpus* Petuch, 1988 (the true genus *Turritella* is confined to the southwestern Pacific and is not found in the Americas)

Naticidae
- *Euspirella* Petuch and Drolshagen, new genus (see Systematic Appendix)
- *Poliniciella* Petuch, 1988

FIGURE 1.15 Detail of a shell bed exposed in a cliff on Urbanna Creek, 4 miles upriver from Urbanna, Middlesex County, Virginia. Here, the assemblage is dominated by the large scallop *Chesapecten middlesexensis* (Mansfield, 1936) and the bittersweet clam *Costaglycymeris mixoni* Ward, 1992. The large, oval, dark-colored brachiopod *Discinisca lugubris* (Conrad, 1834) can be seen to be attached to a fragment of scallop shell in the left center. This type of molluscan assemblage is typical of the Cobham Bay Member of the Eastover Formation.

Muricidae

Chesathais Petuch, 1988
Chesatrophon Petuch, 1988
Lirosoma Conrad, 1862
Mariasalpinx Petuch, 1988
Patuxentrophon Petuch, 1992
Scalaspira Conrad, 1862
Stephanosalpinx Petuch, 1988

Fasciolariidae

Conradconfusus Snyder, 2000
Mariafusus Petuch, 1988
Pseudaptyxis Petuch, 1988

Busyconidae

Sycopsis Conrad, 1867

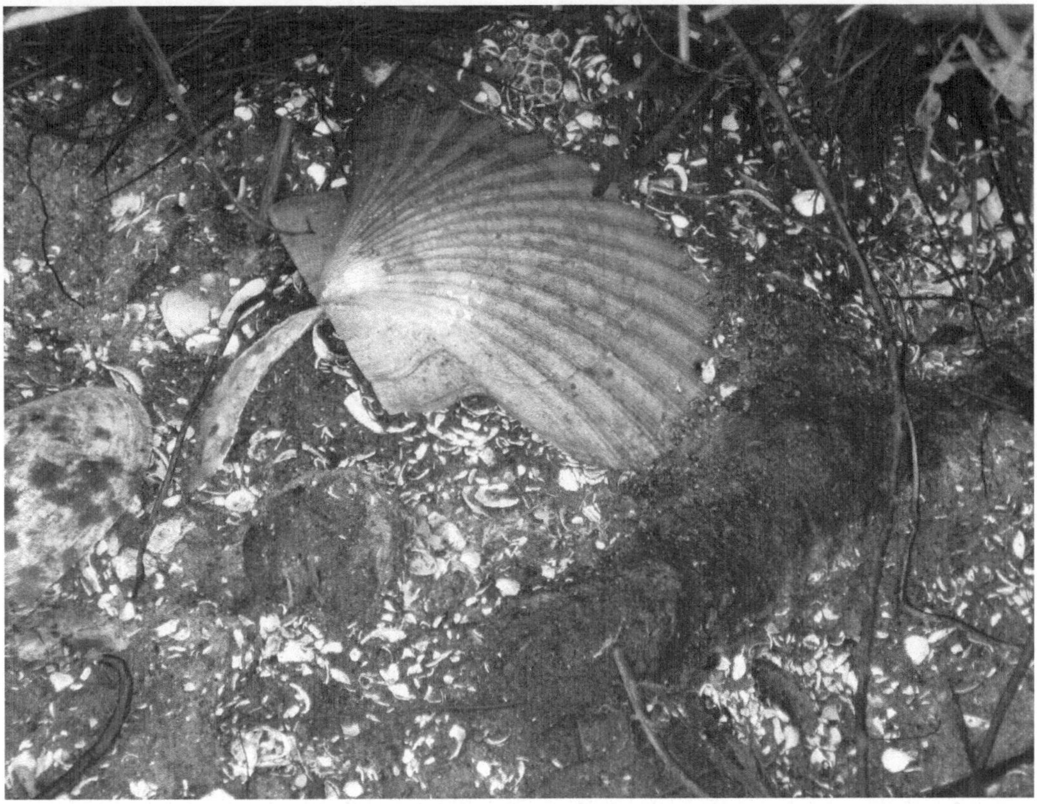

FIGURE 1.16 Detail of an exposure of the Cobham Bay Member of the Eastover Formation in a cliff on Urbanna Creek, 4 miles from Urbanna, Middlesex County, Virginia. Here, the large scallop *Chesapecten middlesexensis* (Mansfield, 1936) can be seen to be embedded in a matrix of *Mulinia rappahannockensis* (Gardner, 1944) marsh clams and fragments of *Costaglycymeris* and *Mercenaria* clams.

Buccinidae
Bulliopsis Conrad, 1862

Volutidae
Megaptygma Conrad, 1863
Scaphellopsis Petuch, 1988

Cancellariidae
Cancellariella Martin, 1904
Marianarona Petuch, 1988
Mariasveltia Petuch, 1988

Terebridae
Laevihastula Petuch, 1988

Turridae
Calverturris Petuch, 1992

Chesaclava Petuch, 1988
Chesasyrinx Petuch, 1988
Mariadrillia Petuch, 1988
Mariaturricula Petuch, 1988
Nodisurculina Petuch, 1988
Sediliopsis Petuch, 1988
Transmariaturris Petuch, 1988

The tropical Baitoan Province, on the other hand, contained a much richer gastropod fauna with entire suites of genera and families that were missing from the Transmarian Province. Some of these missing classic Tethyan tropical gastropod families, richly represented in the Baitoan region, include the Turbinidae (both subfamilies Turbinidae and Astraeinae), Potamididae, Cerithiidae, Modulidae, Strombidae, Cypraeidae, Ovulidae, Triviidae, Melongenidae, the volutid subfamilies Lyriinae and Volutinae, and the Turbinellidae (both subfamilies Turbinellinae and Vasinae) (Petuch, 2004:21–56).

All four subseas of the Salisbury Sea, the Old Church, Calvert, Patuxent, and St. Mary's, were included within the Transmarian Province. The Albemarle Sea, on the other hand, was split between three separate faunal provinces, reflecting the shifting current structure along the Carolina and southern Virginia coastlines. The Chattian Oligocene and Aquitanian Miocene Silverdale Subsea of the Albemarle Sea was included in the Baitoan Province (as the Onslowian Subprovince; Petuch, 2004), while the Pliocene Williamsburg, Yorktown, and Croatan Subseas of the Albemarle Sea were included in the subsequent Pliocene–Pleistocene Caloosahatchian Province (Petuch, 2004). Only the Burdigalian–Langhian Pamlico Subsea of the Albemarle Sea and the late Tortonian–Messinian Rappahannock Subsea of the Albemarle Sea were included in the Transmarian Province. During the Burdigalian and Langhian, when the Salisbury and Albemarle Seas were separated by the Norfolk Peninsula, the Salisbury Sea (Calvert Subsea) constituted the **Calvertian Subprovince** of the Transmarian Province, and the Albemarle Sea (Pamlico Subsea) constituted the **Pungoian Subprovince** of the Transmarain Province.

The last remnant of the Transmarian Province existed in the Rappahannock Subsea of the Albemarle Sea (fossils of the Eastover Formation), and the province disappeared after the "Messinian Event." Some of the last-living Transmarian elements, which make their final appearance in the Rappahannock Subsea, include the gastropods *Mariasalpinx* (Muricidae), *Bulliopsis* (Buccinidae), and *Laevihastula* (Terebridae). A few classic Transmarian genera did manage to survive the "Messinian Event" and continued on into the early Pliocene Williamsburg and Yorktown Subseas of the Albemarle Sea (found in the Sunken Meadow and Rushmere Members of the Yorktown Formation). Some of these include the gastropods *Chesatrophon, Scalaspira,* and *Lirosoma* (Muricidae), *Mariafusus* (Fasciolariidae), *Scaphellopsis* and *Megaptygma* (Volutidae), and *Sediliopsis* (Turridae) and the bivalve genus *Glossus* (Glossidae). These early Pliocene Transmarian relicts lived only in the Yorktown Subsea and did not occur farther south in the Carolinas or Florida. By the mid-Pliocene, these taxa also became extinct.

2 Geologic Framework of the Maryland and Virginia Miocene

The geological formations of coastal Maryland and Virginia, in reality the seafloors deposited within the Salisbury and Albemarle basins, are placed together in the **Chesapeake Group** (Clark, 1904:29–30). Within this group, eight formations are recognized: the Oligocene–Miocene **Old Church Formation**, the Miocene **Calvert**, **Choptank**, **St. Mary's**, and **Eastover Formations**, the Pliocene **Yorktown** and **Chowan River Formations**, and the early Pleistocene **James City Formation** (Figure 2.1 and Figure 2.2). All eight formations have similar lithologies, being composed of varying amounts of clay, mud, and sand mixed with thick beds of fossil shells. The molluscan fossil component contributes a high carbonate concentration to these formations, setting aside the Chesapeake Group from the other geological units within the Atlantic Coastal Plain. In this book, only the four Miocene formations of the Chesapeake Group will be discussed.

Although often appearing very similar in composition, the Miocene formations and members of the Chesapeake Group are separated by readily distinguishable unconformities and represent separate depositional events (Ward, 1992). During the regressional times of the unconformities (see Chapter 1), intense ecological pressures caused by deteriorating climates resulted in the extinction of many taxa and the rapid evolution of the survivors. Many of these rapidly evolving genera found within the Miocene shell beds of the Chesapeake Group, such as the bivalves *Chesapecten, Marvacrassatella,* and *Mercenaria,* and the gastropod *Ecphora,* can be used as geochronological guides for determining the formational and member boundaries. In Chapters 4, 5, 6, and 7 of this book, we illustrate over 260 molluscan index fossils and list over 450 species, all arranged by formation and member. As mentioned in Chapter 1, the late Oligocene–early Miocene Old Church Formation will not be covered in this book, due to its having few natural exposures and poorly preserved fossils.

In 1904, George B. Shattuck published the first workable stratigraphic scheme for the Chesapeake Miocene. In this work, he proposed a system of 24 stratigraphic units ("Zones") and formally described, for the first time, the lithologies of the Calvert, Choptank, and St. Mary's Formations. In Shattuck's scheme, Zones 1 through 15 encompassed the Calvert Formation, Zones 16 through 20 encompassed the Choptank Formation, and Zones 21 through 24 encompassed the St. Mary's Formation (Shattuck, 1904:33–137). Two new members of the Calvert Formation were proposed in this work: the **Fairhaven Member** (originally "Fairhaven Diatomaceous Earth," Zones 1 through 3) and the **Plum Point Member** (originally "Plum Point Marls," Zones 4 through 15). Interestingly enough, no member names were proposed for the Choptank and St. Mary's Formations, although distinct lithological differences occur. Shattuck's scheme, although modified by subsequent workers, is still a useful stratigraphic classification system and will be used throughout this book. Since the formations dip to the south in Chesapeake Bay, it is possible to follow any

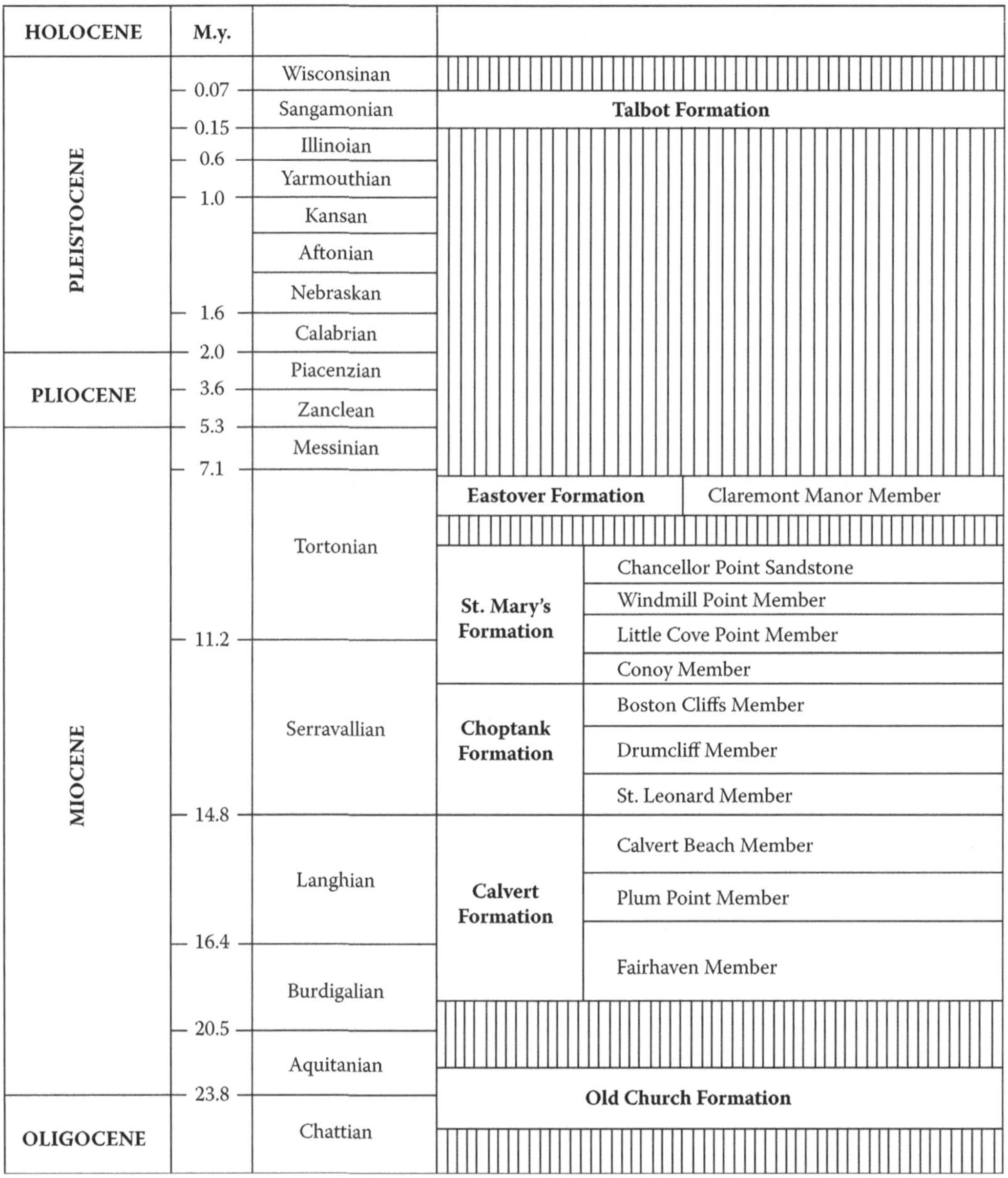

FIGURE 2.1 Generalized stratigraphic column for the Chesapeake Bay area of Maryland, showing the latest Paleogene and Neogene formations and major unconformities. This column includes the areas of the Calvert Cliffs and the Patuxent, Potomac, and St. Mary's Rivers. The entire Pliocene and most of the Pleistocene can be seen to be missing.

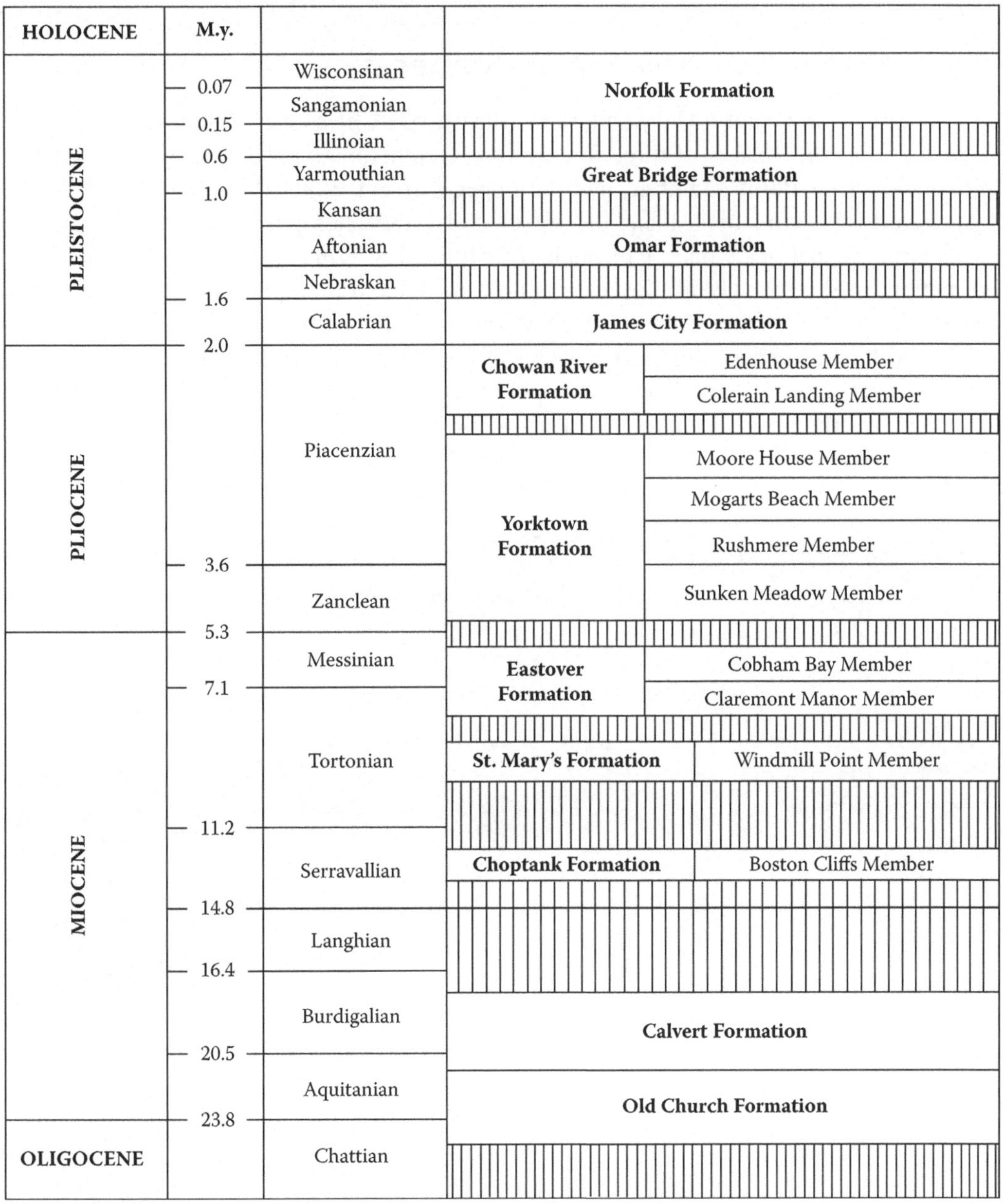

FIGURE 2.2 Generalized stratigraphic column for northern Virginia, showing the latest Paleogene and Neogene formations and major unconformities. This column includes the Chesapeake Bay area of Virginia and the areas of the Potomac, Rappahannock, Mattaponi, Pamunkey, York, and James Rivers.

Shattuck Zone from its first appearance high on the Calvert Cliffs to where it disappears below the beach. A fossil collector only needs to walk south on the beach in order to intersect any of the Shattuck Zones.

THE CALVERT FORMATION AND ITS MEMBERS

Named for Calvert County, Maryland, site of the Calvert Cliffs (over 30 miles in length), the Calvert Formation consists of three fossiliferous members: the lower **Fairhaven Member**, the middle **Plum Point Member**, and the upper **Calvert Beach Member** (Ward, 1992). The Calvert is also the thickest and best developed of the Chesapeake Miocene formations, encompassing Shattuck Zones 1 through 16. Although originally placed in the Choptank Formation by Shattuck (1904), Zone 16 is now considered the uppermost bed of both the Calvert Formation and the Calvert Beach Member (Ward, 1992).

The Fairhaven Member (Shattuck Zones 1 through 3; named for Fairhaven, southernmost Anne Arundel County, Maryland) is composed mostly of a hard, blue, diatomaceous clay, containing poorly preserved fossils. Shattuck Zone 2, at some localities, contains thin layers of white sand with rich shell deposits, but these are only locally developed. The uppermost bed, Shattuck Zone 3 (Figure 2.3), is the thickest and most extensive Fairhaven unit, cropping

FIGURE 2.3 View of some of the northernmost Calvert Cliffs, 2 miles south of Chesapeake Beach, Calvert County, Maryland. Here, the blue-gray diatomaceous clays of the Fairhaven Member (Shattuck Zone 3), Calvert Formation, can be seen cropping out at the base of the low cliff, along the tide line. The rare ecphora *Trisecphora chamnessi* (Petuch, 1989) is occasionally collected at this locality.

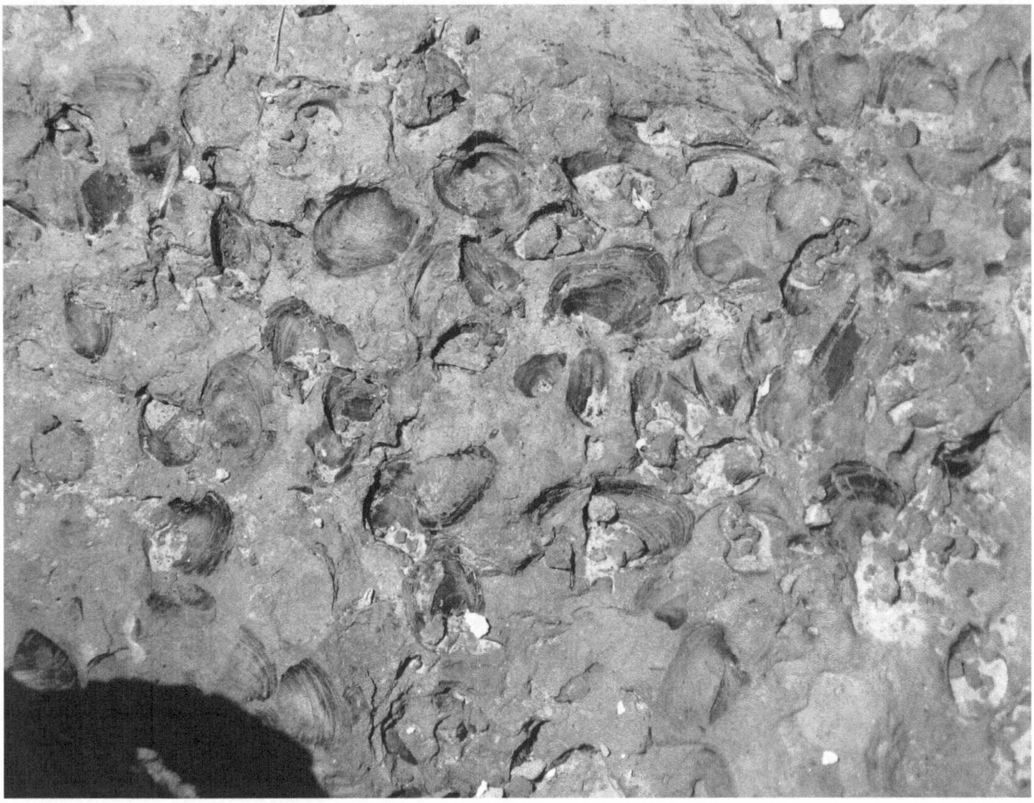

FIGURE 2.4 Detail of Shattuck Zone 3 of the Fairhaven Member, Calvert Formation, in the northernmost Calvert Cliffs 2 miles south of Chesapeake Beach, Calvert County, Maryland. Here, the imprints of the glossid clam *Glossus mazlea* subspecies can be seen in the hard diatomaceous clay of the exposed cliff face.

out near the tide line for several miles around Chesapeake Beach, northernmost Calvert County. This hard clay layer often contains impressions of the bivalves *Lucinoma contracta* (Say, 1824) and *Glossus mazlea* subspecies (Figure 2.4) and crushed, broken specimens of the rare ecphora *Trisecphora chamnessi* (Petuch, 1989) (see Chapter 4).

The Plum Point Member (Shattuck Zones 4 through 13; named for Plum Point, Calvert County) is the thickest member of the Calvert Formation and contains the richest and best-preserved shell beds. A particularly interesting and characteristic Plum Point bed is seen in Shattuck Zone 4, which contains a thick monoculture layer of the gryphaeid oyster *Hyotissa percrassa* (Conrad, 1840) (Figure 2.5). Of primary interest in the Plum Point Member is Shattuck Zone 10, a thick sand and clay bed that contains the most diverse molluscan fauna (see Chapter 4). Zone 10 extends from Chesapeake Beach, where it is high up on the cliff face (Figure 2.6), to approximately 3 miles south of Plum Point, Calvert County, where it intersects the tide line. Here the molluscan assemblage is dominated by the scallop *Chesapecten coccymelus* (Dall, 1898) and several rare and unusual ecphoras (see Chapters 2, 3, and 4).

The sands of the Calvert Beach Member (Shattuck Zones 14 through 16; named for Calvert Beach, Calvert County) are heavily leached, and its fossils are poorly preserved. Only thick calcitic species, such as ecphoras, are well preserved enough to be collectible.

FIGURE 2.5 View of the Calvert Cliffs, 5 miles south of Chesapeake Beach, Calvert County, Maryland. Here, a densely packed layer of the gryphaeid oyster *Hyotissa percrassa* (Conrad, 1840) can be seen in Shattuck Zone 4 of the Plum Point Member, Calvert Formation (exposed at the base of the cliff). The beach is littered with tiny *Caryocorbula elevata* (Conrad, 1838) valves, washed from Shattuck Zones 5, 6, and 7 in the cliff face above the Shattuck Zone 4 *Hyotissa* bed.

Throughout its extent, from 3 miles south of Plum Point to south of Flag Pond, Calvert County, the main beds of the Calvert Beach Member (Shattuck Zones 14 and 16) contain some of the rarest and most unusual ecphora species found anywhere along the Calvert Cliffs (see Chapter 4). Several Zone 16 localities north of Flag Pond have yielded beds of porpoise coprolites (fossilized dung), complete with component fish bones.

THE CHOPTANK FORMATION AND ITS MEMBERS

Named for the Choptank River of the Delmarva Peninsula, the Choptank Formation was originally described without members (Shattuck, 1904). The formation was also originally larger in scope, including Shattuck Zone 16 (now placed in the Calvert Formation). Subsequent workers (primarily Ward, 1992) have clarified the true structure of the Choptank Formation by removing Shattuck Zone 16 and describing three new members. These new subdivisions of the Choptank were based on depositional cycles and unconformities and include the lower **St. Leonard Member** (Shattuck Zone 17), the middle **Drumcliff Member** (Shattuck Zone 18), and the upper **Boston Cliffs Member** (Shattuck Zone 19).

FIGURE 2.6 Detail of the upper cliff face along the Calvert Cliffs, 6 miles south of Chesapeake Beach, Calvert County, Maryland. Here, the fossil-rich Shattuck Zone 10 of the Plum Point Member, Calvert Formation can be seen intercalated between the unfossiliferous Shattuck Zones 9 and 11. At this locality, the small bivalve *Astarte cuneiformis* Conrad, 1840 is abundant and makes up a large percentage of the shells exposed in the cliff.

The St. Leonard Member (named for St. Leonard, Calvert County, Maryland) is particularly well developed along the Calvert Cliffs near St. Leonard and Flag Pond (Figure 2.7). Here, it is typically a reddish-yellow sand filled with one of the most interesting faunas of the Chesapeake Miocene (see Chapter 5). The giant pearly oyster *Isognomon (Hippochaeta) torta* (Say, 1820) and the branching ahermatypic coral *Astrhelia palmata* (Goldfuss, 1826) dominate the ecosystems of this member, often producing large biohermal, reeflike structures. These are the only "coral reefs" of the Chesapeake Miocene, and they housed a number of rare gastropod mollusks, including the muricids *Stephanosalpinx candelabra* Petuch, 1988 and *Ecphorosycon lindajoyceae* Petuch, 1993. The volute shell *Volutifusus conradianus* (Martin, 1904) is also common in the St. Leonard Member, often occurring in large aggregations (see Chapter 5).

The largest and most diverse molluscan fauna is found in the Drumcliff Member (named for Drumcliff, on the Patuxent River, St. Mary's County, Maryland), which is best developed along the south shore of the Patuxent River. Here, large beds of the scallop *Chesapecten nephrens* Ward and Blackwelder, 1975 dominated the benthic ecosystem. These scallop beds also predominate in the Drumcliff Member along the Calvert Cliffs near Matoaka

FIGURE 2.7 View of the Calvert Cliffs north of Matoaka Cottages, St. Leonard, Calvert County, Maryland. Here, the St. Leonard Member (Shattuck Zone 17) of the Choptank Formation can be seen exposed at the base of the cliff (on the right). Bioherms of the coral *Astrhelia palmata* (Goldfuss, 1826) and the giant pearly oyster *Isognomon (Hippochaeta) torta* (Say, 1820) are frequently encountered along this area. The sharp contact with the Drumcliff Member (Shattuck Zone 18) can be seen at the top of the St. Leonard Member. A large landslide, typical of the eroding Calvert Cliffs, can be seen to have covered the Choptank Formation farther down the beach.

Cottages, St. Leonard, Calvert County. Along the Patuxent River and its tributaries, the Drumcliff Member is usually a reddish-yellow sand packed with fossil shells, closely resembling the underlying St. Leonard Member. At other areas, such as south of Governor Run and the St. Leonard area of the Calvert Cliffs, the Drumcliff is composed of a bluish sandy clay, often devoid of fossils (Figure 2.7).

The Boston Cliffs Member (named for the Boston Cliffs on the Choptank River, eastern shore of Maryland) lithologically resembles the underlying Choptank members, being composed of a reddish-yellow sand and thickly packed beds of fossil shells. This uppermost Choptank member contains a much more impoverished and less diverse molluscan fauna than does the underlying Drumcliff Member, and this difference in faunas can be used to differentiate the two members. The small, rare scallop *Chesapecten monicae* Ward, 1992 (see Chapter 5) is confined to the Boston Cliffs Member and is most frequently encountered near Flag Pond, Calvert Cliffs.

THE ST. MARY'S FORMATION AND ITS MEMBERS

Named for St. Mary's County, Maryland (Shattuck, 1904:72), the St. Mary's Formation crops out along the southern Calvert Cliffs, the southern shore of the Patuxent River, the

St. Mary's River, and along the headwaters of the Rappahannock and Mattaponi Rivers in Essex, King William, and Richmond Counties in northern Virginia. As now understood, the St. Mary's contains three members: the lower **Conoy Member** (Shattuck Zone 20), the middle **Little Cove Point Member** (Shattuck Zones 21, 22, and 23), and the **Windmill Point Member** (Shattuck Zone 24) (Ward, 1992). Although originally included in the Choptank Formation as its uppermost unit (Gernant, 1970; Shattuck, 1904:72), the Conoy Member has subsequently been placed in the St. Mary's Formation as its lowermost unit (Ward, 1992). A possible unnamed fourth member has been recognized in a small stretch of the St. Mary's River, from south of Chancellor Point to St. Inigoe's Creek (Petuch, 1988, 2004). This possible new member will be referred to as the **Chancellor Point Sandstone** in this book.

The Conoy Member (named for Camp Conoy, Calvert County, Maryland) consists of a green-blue sand mottled with red patches and bands of oxidized iron. In some localities, the entire exposed surface is weathered to a rusty-red color. The Conoy has a limited area of outcrops, extending from only near Flag Pond to north of Little Cove Point, Calvert County. Throughout its range, the Conoy contains few fossils, mostly badly preserved specimens of the turritellid *Mariacolpus* and fragments of the scallop *Chesapecten.* The only well-preserved index fossil for the Conoy Member is the muricid gastropod *Ecphora conoyensis* Petuch, 2004 (see Chapter 6, Figure 6.1), which is most often collected north of Little Cove Point, southern Calvert County. This small, heavily constructed ecphora is the perfect morphological intermediate, both in shape and size, between the Choptank (Boston Cliffs Member) *Ecphora williamsi* Ward and Gilinsky, 1988 and the St. Mary's (lower Little Cove Point Member) *Ecphora asheri* Petuch, 1988.

The Little Cove Point Member (named for Little Cove Point, southern Calvert County, Maryland) is lithologically the most complex of the St. Mary's members, being composed of three distinct beds (Shattuck Zones 21, 22, and 23). All three units contain well-preserved fossil mollusks, although Shattuck Zone 22 contains the richest and best-preserved fossil beds. The unconformity between Shattuck Zones 21 and 22 appears to be larger than that between Shattuck Zones 22 and 23, because Shattuck Zone 21 has a different set of index species than do Shattuck Zones 22 and 23. This implies a greater period of missing time between Shattuck Zones 21 and 22. The Little Cove Point Member is best exposed along the stretch of Calvert Cliffs between Flag Pond and Drum Point near Solomons, Calvert County.

Although not as thick and well-developed as the underlying Little Cove Point Member, the Windmill Point Member (Shattuck Zone 24; named for Windmill Point, St. Mary's County, Maryland) contains the richest molluscan fauna and the best-developed shell beds of the St. Mary's Formation (see Chapter 6). For this reason, the Windmill Point contains the highest carbonate fraction of the entire St. Mary's Formation and is among the most carbonate-rich units in the Miocene component of the Chesapeake Group. In some areas along the St. Mary's River, such as at Chancellor Point, the Windmill Point Member contains dense beds of the turritellid gastropod *Mariacolpus lindajoyceae* Petuch, 2004 that are so closely packed that little clay matrix is present between individual specimens. The Windmill Point contains a different set of index fossils than does Shattuck Zone 23, indicating that the intervening unconformity represents a period of time long enough to allow the evolution of an entirely new fauna.

Along the St. Mary's River, from St. Inigoe's Creek to Chancellor Point, a thick (0.5–1 m) ledge of dense olive-gray sandstone can be seen projecting above the shelly clays of the

Windmill Point Member. In some places, the ledge is undercut as much as 1 m, producing small, fossil-filled caves in the Windmill Point. Since the ledge is located along the tide line, it is often the only traversable area at high tide. Over the last two decades, however, much of the original exposed sandstone ledge has been covered by riprap and piers, the result of housing development along the St. Mary's River. Small exposures are still found at the base of the low cliffs just south of Chancellor Point and at the mouth of St. Inigoe's Creek. Because of its unique lithology, being composed of a dense, heavily indurated fossiliferous olive-gray sandstone, this layer has been referred to as the Chancellor Point Sandstone (named for Chancellor Point on the St. Mary's River, St. Mary's County, Maryland) (see Petuch, 2004).

Although Ward (1992:33) was the first to describe the lithology of the sandstone unit, the characteristic Chancellor Point Sandstone molluscan fauna was only recently recognized and described by the senior author (1997, 2004). Even though most of the fossil shells in the Chancellor Point Sandstone are fragmentary and poorly preserved, enough complete specimens have been collected to show that the fauna is significantly different from that of the underlying Windmill Point Member (Shattuck Zone 24) (see Chapter 6). This demonstrates that a sizable unconformity exists between the Chancellor Point Sandstone and the underlying Windmill Point Member, probably equivalent to that between the Windmill Point and Little Cove Point Members. The small areal extent and relative thinness of the Chancellor Point Sandstone indicate that this distinctive sandstone unit represents the erosional remnant of a once much thicker and more extensive unit. South of Chancellor Point, near St. Inigoe's Creek, a thick (2 m) unfossiliferous blue clay unit overlies the Chancellor Point Sandstone. Containing large gypsum crystals, this clay unit probably dates from the late Tortonian and may represent the lower part of the Claremont Manor Member of the Eastover Formation (see Chapter 7).

THE EASTOVER FORMATION AND ITS MEMBERS

Although formally described and named relatively recently (Ward and Blackwelder, 1980), the Eastover Formation (named for Eastover, Surry County, Virginia) was long known as the "Virginia St. Mary's" (Gardner, 1948; Mansfield, 1928). This misrepresentation was due primarily to the similarities of the lithology and molluscan paleontology to those of the St. Mary's Formation. Later research (Petuch, 2004; Ward, 1992; Ward and Blackwelder, 1980) has shown that the molluscan faunas of the two formations are completely different and that the Eastover represents a much younger unit. The Eastover Formation extends from the northern shoreline of the Potomac River and southern St. Mary's River of Maryland, across Virginia, inland to Richmond County and to the Virginia Chesapeake shoreline, and south to northern North Carolina. Although Ward and Blackwelder (1980:D25) state that the Eastover is missing in the phosphate mines at Aurora, North Carolina, subsequent work has shown that the formation is indeed present, although patchily distributed and eroded by paleoriver channels (Petuch, 2004:123, 126).

The Eastover Formation contains two members: the lower **Claremont Manor Member** (late Tortonian Age) and the upper **Cobham Bay Member** (Messinian Age). Both members are lithologically similar in appearance and are separated by a distinct unconformity.

FIGURE 2.8 Detail of an exposure of the Claremont Manor Member of the Eastover Formation, along the headwaters of Urbanna Creek, 4 miles from Urbanna, Middlesex County, Virginia. Here, exposed near the high tide line, the Claremont Manor sediments can be seen to be filled with bivalves such as the large scallop *Chesapecten ceccae* Ward, 1992 (specimens on either side of the hammer handle), the bittersweet clam *Costaglycymeris virginiae* (Dall, 1898) (specimens at the base of the hammer), and the giant pearly oyster *Isognomon (Hippochaeta) sargenti* Petuch and Drolshagen, new species (fragments in the layer above the hammer).

As a result of this period of missing time, the Claremont Manor and Cobham Bay molluscan faunas are quite different, with only a few species in common. The molluscan faunas of both members are still relatively unexplored, and many new species are yet to be discovered and named.

The Claremont Manor Member (named for Claremont Manor on the James River, Surry County, Virginia) is composed of a bluish-gray clay that weathers to a reddish-tan. Having been deposited in cooler water conditions (see Chapter 1), the Claremont Manor contains a less-diverse molluscan fauna than does the subsequent Cobham Bay. In the headwaters of Urbanna Creek, Urbanna, Middlesex County, Virginia (Figure 2.8), the Claremont Manor crops out at the low tide line and contains a higher quartz sand content than does the stratotype locality on the James River. Here, the large scallop *Chesapecten ceccae* Ward, 1992 dominates the molluscan assemblage, along with the large bivalves *Costaglycymeris virginiae* (Dall, 1898), *Marvacrassatella surryensis* (Mansfield, 1929), and *Mercenaria druidi gardnerae* Petuch and Drolshagen, new subspecies (see Systematic Appendix). These, together with the giant pearly oyster *Isognomon (Hippochaeta) sargenti* Petuch

FIGURE 2.9 Detail of an exposure of the Cobham Bay Member of the Eastover Formation, along the eastern side of Urbanna Creek, 2 miles from Urbanna, Middlesex County, Virginia. Here, a thick, densely packed layer of the small marsh clam *Mulinia rappahannockensis* (Gardner, 1944) can be seen to be overlain by a layer of shell fragments, mostly of the larger bivalves *Costaglycymeris mixoni* Ward, 1992, *Marvacrassatella urbannaensis* (Mansfield, 1929), and *Mercenaria druidi* Ward. 1992.

and Drolshagen, new species (see Systematic Appendix), make up the bulk of the carbonate fraction.

Also composed of a bluish-gray clay that weathers to a reddish-tan, the Cobham Bay Member (named for Cobham Bay on the James River, Surry County, Virginia) differs from the underlying Claremont Manor in having a higher quartz sand content and in having more complex sets of lithologies. These are often manifested as alternating beds of shell hash, shelly sand, large shell fragments, and sandy clay (Figure 2.9). As pointed out in Chapter 1, a chain of low islands, the Bayport Archipelago, bisected the Rappahannock Subsea during Cobham Bay time. In the protected estuarine areas to the west of the archipelago, muds, clays, and fine sands were deposited. On the eastern side of the island chain, in more energetic, open oceanic conditions, coarse sand and shell hash were deposited. The clays of the protected estuarine lagoons housed immense aggregations of the small marsh clam *Mulinia rappahannockensis* (Gardner, 1944), while the sand and shell hash bottoms along the open ocean side housed huge

FIGURE 2.10 Detail of an exposure of the Cobham Bay Member of the Eastover Formation, along the eastern side of Urbanna Creek, 3 miles from Urbanna, Middlesex County, Virginia. Here, the giant pearly oyster *Isognomon (Hippochaeta) sargenti* Petuch and Drolshagen, new species grew together in a giant biohermal, reeflike assemblage and formed the substrate for a large number of epizoonts, including the keyhole limpet *Glyphis lindae* Petuch and Drolshagen, new species, the slipper shells *Crepidula (Bostrycapulus) bonnieae* Petuch and Drolshagen, new species, and *Crepidula virginica* Conrad, 1871, and the flattened discinid brachiopod *Discinisca lugubris* (Conrad, 1834).

beds of the scallop *Chesapecten middlesexensis* (Mansfield, 1936) and the giant pearly oyster *Isognomon (Hippochaeta) sargenti* (Figure 2.10). The disappearance of this distinctive pearly oyster can be used to separate the Miocene Eastover Formation from the similar-appearing Pliocene Yorktown Formation (Ward and Blackwelder, 1980:D11) (see Chapter 7).

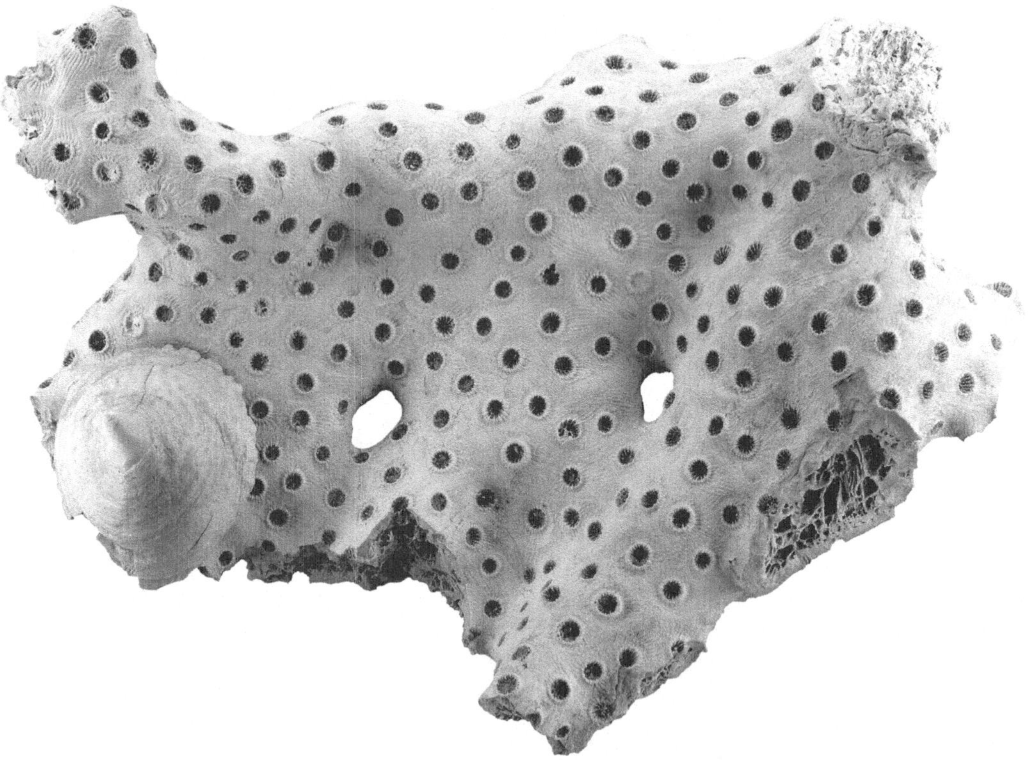

A specimen of the cup-and-saucer shell *Crucibulum multilineatum* (Conrad, 1842) attached to the base of an *Astrhelia palmata* (Goldfuss, 1826) coral colony (coral length 120 mm). Coral bioherms ("reefs") of *Astrhelia palmata* are typical of the St. Leonard Member of the Choptank Formation.

3 Ecphoras of the Maryland and Virginia Miocene

In the Miocene shell beds of Maryland and Virginia, the single most spectacular species radiation is seen in the ecphora shells. These bizarrely shaped gastropods (Family Muricidae, Subfamily Ocenebrinae), with their distinctive ribbed shell sculpture, are collectively considered the classic index fossil group for the Chesapeake Miocene. Because of their unusual shell sculpture, large size, and general intrinsic beauty, ecphoras are also highly prized and sought-after by shell collectors and amateur paleontologists and rank second only to shark teeth as the most desirable fossils in the Chesapeake Bay area. The early colonial settlers also found ecphoras intriguing, as a specimen of the genus *Ecphora* was the first American fossil gastropod illustrated in a European natural history text (in the *Historiae sive Synopsis Methodique Conchiliorum* of Martin Lister, the 1770 Huddlesford edition).

The higher taxa of the ecphoras can be separated on the basis of four main aspects of shell morphology:

1. **Shell Shape**: being either inflated and globose or narrow and cylindrical, and having either a consistently high or low spire
2. **Number of Ribs**: with shells consistently having three, four, five, or multiple main ribs and, in some cases, multiple secondary ribs
3. **Rib Form**: with the ribs being consistently either thin, narrow and bladelike, or wide and rounded, or in some cases, extremely wide and "T"-shaped in cross section
4. **Development of the Siphonal Canal**: with some groups having long, broad canals with flaring, wide umbilici, and other groups with long, narrow canals and slender umbilici, and still other groups with short, stubby canals and umbilici that are virtually closed

With the exception of the most primitive forms, all ecphora shells have a thick calcite outer shell layer, often colored bright reddish-brown or yellowish-tan. This resistant outer shell layer prevents dissolution by groundwater, often making ecphoras some of the only fossils to survive in heavily leached deposits (such as Shattuck Zones 12 and 14 of the Calvert Formation and Shattuck Zone 20 of the St. Mary's Formation).

The ecphora fauna of Maryland and Virginia comprises 40 named species belonging to four genera and one subgenus. These generic groups include ***Ecphora*** Conrad, 1843—medium to large-sized shells with four raised ribs and with the ribs being "T"-shaped in cross section, with either slightly or broadly flattened edges, and with a wide, open umbilicus; type species *Ecphora quadricostata* (Say, 1824), Yorktown Formation of Virginia (examples on Figure 3.1); ***Planecphora*** Petuch, 2004—originally proposed as a full genus and now considered a subgenus; small to medium-sized shells with four raised, very thin, bladelike ribs with rounded, not flattened, edges, and with a wide, open umbilicus; type species *Ecphora choptankensis* Petuch, 1988 (examples on Figure 3.1); ***Trisecphora*** Petuch,

A
B
C
D
E
F
G
H
I
J
K

1988a—originally proposed as a subgenus and now considered a full genus; medium to small-sized shells with three raised ribs, and with the ribs being either straight and thin with rounded edges, cupped and posteriorly recurved, or broadly flattened and "T"-shaped in cross section; type species *Ecphora tricostata* Martin, 1904 (examples on Figure 3.2); ***Chesathais*** Petuch, 1988a—high-spired, fusiform shells with multiple low, rounded, cord-like ribs, usually 5 to 13 in number, and with a narrow umbilicus that is almost closed; type species *Chesathais lindae* Petuch, 1988 (Examples on Figure 3.2); ***Ecphorosycon*** Petuch, 1988a—large shells with three or four low, thin cords, and with the edge of the shoulder extending posteriorly, forming a broad, bladelike shelf; type species *Ecphora pamlico* Wilson, 1987 (Examples on Figure 3.2). The genera *Ecphora, Trisecphora,* and *Ecphora (Planecphora)* underwent extensive species radiations during Calvert and Choptank times (Burdigalian to Serravallian Miocene), producing the bulk of the ecphora species encountered along the Calvert Cliffs. The genera *Ecphorosycon* and *Chesathais* were far less species rich, being represented by only a single species in any stratigraphic unit.

Prior to Calvert time, from the Oligocene to the early Burdigalian Miocene, two other ecphora genera existed: the long-siphoned, busycon-shaped *Siphoecphora* Petuch, 1989 and the small, short-siphoned *Rapanecphora* Petuch, 1989 (Petuch, 1989, 2004). *Siphoecphora* (typified by *S. auroraensis* (Wilson, 1987)) is found only in the Burdigalian–Langhian Pungo River Formation of North Carolina, although broken fragments from Shattuck Zone 2 of the Fairhaven Member (Calvert Formation) may represent Maryland specimens of this species. The genus *Rapanecphora* (typified by *R. vaughani* (Mansfield, 1937)) is restricted to the Chattian Oligocene and Aquitanian Miocene of Florida and the Carolinas and represents the oldest ecphora lineage.

After Eastover time (Messinian Miocene), the ecphoras lived on into the Pliocene and underwent a series of species radiations in Florida (Murdock Station and Tamiami Formations), the Carolinas (Duplin Formation), and Virginia (Yorktown Formation). *Ecphora* (s.s.) and *Planecphora* survived until the mid-Piacenzian Pliocene and evolved several species and subspecies. These coexisted with two other ecphora groups, the Pliocene-restricted *Latecphora* Petuch, 1989 (with four extremely wide, "T"-shaped ribs that almost fuse together; typified by *L. bradleyae* (Petuch, 1987)) and the late Miocene–Pliocene *Globecphora* Petuch, 1994 (highly inflated globose shells with thin, almost obsolete cords; typified by *G. floridana* (Petuch, 1989)). By the end of the mid-Piacenzian Pliocene (3 million years BP), the last-living group of ecphoras, the genus *Latecphora,* became extinct, ending a lineage of ocenebrine muricids that spanned over 23 million years (late Oligocene–late Pliocene) (see Petuch, 1989; 2004; 2007; 2008).

FIGURE 3.1 **(Facing page)** Examples of ecphora genera and species complexes from the Chesapeake Miocene. **Representatives of the typical "T"-shaped, wide-ribbed genus *Ecphora* Conrad, 1843 include:** A = *Ecphora gardnerae* Wilson, 1987, length 108 mm; B = *Ecphora williamsi* Ward and Gilinski, 1988, length 57 mm; C = *Ecphora chesapeakensis* Petuch, 1992, length 41 mm; D = *Ecphora calvertensis* Petuch, 1988, length 43 mm. **Representatives of the "T"-shaped, narrower-ribbed "*Ecphora meganae* Species Complex" include:** E = *Ecphora meganae* Ward and Gilinsky, 1988, length 69 mm; F = *Ecphora mattinglyi* Petuch, 2004, length 88 mm. **Representatives of the low-ribbed, fusiform "*Ecphora rikeri* Species Complex" include:** G = *Ecphora rikeri* Petuch, 1988, length 80 mm; H = *Ecphora harasewychi* Petuch, 1989, length 32 mm. **Representatives of the thin-ribbed subgenus *Planecphora* Petuch, 2004, include:** I = *Ecphora (Planecphora) choptankensis* Petuch, 1988, length 62 mm; J = *Ecphora (Planecphora) vokesi* Petuch, 1989, length 62 mm; K = *Ecphora (Planecphora) delicata* Petuch, 1989, length 39 mm.

A
B
C
D
E
F
G
H
I
J
K

EVOLUTIONARY PATTERNS OF THE ECPHORAS

The first recognizable ecphora species appear in the tropical–subtropical waters of Oligocene Florida and the Carolinas, and are typified by species such as *Ecphorosycon tampaensis* (Dall, 1892) and *Ecphorosycon wheeleri* (Rossbach and Carter, 1991). These small, primitive forms moved into the Salisbury Sea by the early Miocene and quickly underwent a series of adaptive radiations. Since the Salisbury Sea at that time contained a rich bivalve fauna but very few gastropod predators, the first ecphora invaders would have had an almost unlimited food supply and very few competitors. Based on the wide range of shell sizes seen in the early Miocene Salisbury ecphoras, they had partitioned themselves ecologically in response to the wide range of sizes of their bivalve prey. This rapid partitioning led to the evolution of at least 33 species during Calvert and Choptank time alone (late Burdigalian–early Serravallian).

The largest radiation of ecphora species is seen in the Calvert Formation, with at least 17 species occurring within the Fairhaven, Plum Point, and Calvert Beach Members. Of these, the upper part of the Plum Point (Shattuck Zones 10, 11, and 12) contains the largest number of species, with eight species in Shattuck Zone 10 and four in Shattuck Zone 12. The Calvert Beach Member (Shattuck Zones 14, 15, and 16) contains a more impoverished fauna, with only two species in Shattuck Zone 14 and three in Shattuck Zone 16 (see Chapter 4). The next largest radiation is seen in the Choptank Formation, with 16 species occurring within its three members. The St. Leonard Member (Shattuck Zone 17), with six species, is the most species-rich unit within the Choptank Formation. The other two members, the Drumcliff (Shattuck Zone 18) and the Boston Cliffs (Shattuck Zone 19), both contain five species. Within the St. Mary's Formation, only five species occur, with one in the Conoy Member (Shattuck Zone 20), two in the Little Cove Point Member (in Shattuck Zones 21 and 22), one in the Windmill Point Member (Shattuck Zone 24), and one in the Chancellor Point Sandstone. Similarly, the Eastover Formation contains only two species, with one in the Claremont Manor Member and one in the Cobham Bay Member. During St. Mary's and Eastover time, unlike Calvert and Choptank time, only one species of ecphora occurred in each stratigraphic unit.

The patterns of evolution and extinction in the ecphoras appear to be linked to ecological competition with other drilling muricid gastropods (Petuch, 1989, 1997). These groups, which include the ocenebrines *Urosalpinx* and *Mariasalpinx,* the trophonines

FIGURE 3.2 **(Facing page)** Examples of ecphora genera and species complexes from the Chesapeake Miocene. **Representatives of the genus *Ecphorosycon* Petuch, 1988 include:** A = *Ecphorosycon pamlico* (Wilson, 1987), length 74 mm; B = *Ecphorosycon kalyx* (Petuch, 1988), length 68 mm. **Representatives of the genus *Chesathais* Petuch, 1988 include:** C = *Chesathais drumcliffensis* Petuch, 1989, length 32 mm; D = *Chesathais ecclesiasticus* (Dall, 1915), length 50 mm. **Representatives of the typical wide, "T"-shaped, three-ribbed genus *Trisecphora* Petuch, 1988 ("*Trisecphora tricostata* Species Complex") include:** E = *Trisecphora tricostata* (Martin, 1904), length 35 mm; F = *Trisecphora tricostata* subspecies (Shattuck Zone 12), length 31 mm. **Representatives of the cupped, upwardly (posteriorly) pointing three-ribbed "*Trisecphora schmidti* Species Complex" include:** G = *Trisecphora schmidti* (Petuch, 1989), length 30 mm; H = *Trisecphora martini* (Petuch, 1988), length 25 mm. **Representatives of the thin-ribbed "*Trisecphora patuxentia* Species Complex" include:** I = *Trisecphora prunicola* (Petuch, 1988), length 27 mm; J = *Trisecphora patuxentia* (Petuch, 1988), length 37 mm; K = *Trisecphora bartoni* Petuch and Drolshagen, new species, length 26 mm.

Chesatrophon, Scalaspira, Lirosoma, and *Patuxentrophon,* the muricine *Stephanosalpinx,* the typhinine *Laevityphis,* and the muricopsinine *Murexiella,* all competed with ecphoras for bivalve prey. During Calvert time, other competing muricids were at a minimum, with only the genera *Patuxentrophon, Murexiella,* and *Laevityphis* being present. This allowed the ecphoras to dominate the drilling niche, resulting in the evolution of the largest number of species. In Choptank time, during the mid-Miocene cooling event and Transmarian Extinction (Petuch, 2004:261–262), the non-ecphora muricid genera virtually disappeared from the Maryland fossil record, being represented by only two small and rare genera, *Stephanosalpinx* and *Patuxentrophon.* This lack of competition allowed the Choptank ecphoras to dominate the drilling niche, with different *Chesathais, Ecphora, Ecphora (Planecphora),* and *Trisecphora* species occurring in each Choptank member. The genus *Ecphorosycon* died out at the beginning of the Choptank Formation (in Shattuck Zone 17), possibly a victim of the severe cooling during Transmarian Extinction time.

In the late Serravallian, conditions began to warm and the Salisbury Sea contained the warmest marine climate in its history. This warming trend appeared to be fatal to the Choptank ecphoras, as the genera *Chesathais* and *Trisecphora* became extinct. The subgenus *Planecphora* became regionally extinct (extirpated), surviving farther south in Florida (Petuch, 2007, 2008). Only the genus *Ecphora* survived into St. Mary's time (late Serravallian–early Tortonian). During this warm time, the Salisbury Sea was invaded by aggressive southern genera such as *Urosalpinx,* and the muricid fauna was repopulated with other Transmarian genera such as *Scalaspira, Lirosoma, Chesatrophon,* and *Laevityphis.* These small but aggressive muricids dominated the small, delicate-shelled bivalve niche, making only large, thick-shelled bivalves available to ecphoras. Because of this, ecphoras were shoved to the ecological fringe, with only one large species being found in any given St. Mary's stratigraphic unit. This pattern was carried on into the Eastover Formation of Virginia, where only two large species occur. A comparison of the rich and diverse ecphora fauna of Shattuck Zone 10 (Calvert Formation) with the lone *Ecphora* species of Shattuck Zone 24 (St. Mary's Formation) readily underscores this ecologically driven extinction pattern.

Phylogenetically, ecphoras derive from the ocenebrine genus *Tritonopsis,* which was prevalent in the Eocene of the Mississippi Sea and the coastal waters of the southeastern United States. After the cold time and extinction period that followed the Chesapeake Bay asteroid impact in the latest Eocene (see Poag, 1999; Poag et al., 2004), *Tritonopsis* disappears and leaves two surviving descendant groups, *Rapanecphora* and *Ecphorosycon.* These were the only ecphora groups in the early and middle Oligocene. By the end of the Oligocene and the Aquitanian Miocene, *Rapanecphora* became extinct, while *Ecphorosycon* evolved three new genera, *Chesathais, Siphoecphora,* and *Trisecphora.* During Langhian time, *Trisecphora* evolved its own new lineage, the true four-ribbed *Ecphora.* This newly evolved genus quickly diversified and, by Calvert Beach time (late Langhian), had produced a separate spin-off lineage, the subgenus *Planecphora.* The genera *Ecphorosycan, Chesathais, Trisecphora, Ecphora,* and *Ecphora (Planecphora)* all coexisted in the late Calvert Subsea and produced the largest ecphora species radiation. By the early Serravallian, the genus *Ecphorosycon* became extinct, leaving only the genera *Ecphora, Ecphora (Planecphora), Trisecphora,* and *Chesathais* in the Patuxent Subsea. This last pulse of ecphora diversity disappeared during the mid-Serravallian, when *Chesathais* and *Trisecphora* became extinct and *Planecphora* became extirpated. Only the genus *Ecphora* survived on into the St. Mary's and Eastover Formations of the Chesapeake Miocene.

ECOLOGY OF THE ECPHORAS

Being carnivorous gastropods, the ecphoras preyed upon the rich and abundant bivalve fauna that characterizes the Maryland and Virginia Miocene. Like their closest living relative, the genus *Forreria* (Family Muricidae, Subfamily Ocenebrinae; found along the California coast), the ecphoras drilled holes through the bivalve shells, using their tooth-covered tongue (radula) as a drill bit. This mechanical drilling and scraping was facilitated by the secretion of acids that weakened the victim's shell. Once a hole was produced, the ecphora would have inserted its long proboscis through the opening and rasped chunks of flesh from the living bivalve (see Figure 3.3).

With an abundance of bivalves of different size classes, ranging from tiny, thin-shelled forms such as *Aligena* and *Sportella* to large, thick-shelled genera such as *Mercenaria* and *Marvacrassatella,* the ecphoras quickly partitioned ecologically. Small species, such as *Trisecphora prunicola* (Figure 3.2I) and *Ecphora (Planecphora) delicata* (Figure 3.1K) most probably preyed upon small, delicate, thin-shelled clams, while large species such as *Ecphora gardnerae* (Figure 3.1A) and *Ecphora rikeri* (Figure 3.1G) were the major predators of large, thick-shelled species. The ecphoras competed heavily with only one other drilling carnivorous gastropod group, the moon snails (Family Naticidae). The drilling technique and types of drill holes differed between the two families; the Naticidae produce a perfectly round, beveled hole that is conical in cross section while the Muricidae produce a round or slightly irregular hole that is straight-sided in cross section. Since there was such an overabundance of bivalve prey, the two groups seemed to have coexisted without catastrophic competition (the Volterra–Gauss Principle). Ecphoras were also cannibalistic, with larger species sometimes drilling and feeding upon smaller species (Figure 3.3B).

The drilling feeding mode of the ecphoras is readily seen in large bivalves collected from the Windmill Point Member of the St. Mary's Formation. Many of the valves of the large venerid bivalve *Mercenaria tetrica* are found with large holes drilled near their hinges (Figure 3.3D). Some of the holes are over 5 mm in diameter and are all of the typical straight-sided muricoidean drilling pattern. With the exception of *Ecphora gardnerae,* all the Windmill Point muricids (*Chesatrophon, Urosalpinx, Mariasalpinx, Lirosoma, Scalaspira,* and *Laevityphis*) are all under 25 mm in length and would have been incapable of drilling 5-mm-diameter holes in *Mercenaria* valves. This leaves only one possibility: *Ecphora gardnerae* must have been the culprit and apparently was the top predator on large bivalves such as *Mercenaria, Dosinia* (Figure 3.3H), *Dallarca,* and *Chesacardium.* Large gastropods such as *Conradconfusus parilis* were also attacked and drilled by *Ecphora gardnerae* (Figure 3.3C). This same ecological pattern is seen in the Eastover Formation of Virginia, where the Claremont Manor *Ecphora whiteoakensis* is seen to have preyed upon *Mercenaria druidi gardnerae* (Figure 3.3A) and *Chesapecten ceccae* (Figure 3.3E).

While living on soft substrates on the open sea floor, and while feeding on shallow infaunal bivalves, ecphoras were themselves vulnerable to molluscivorous predators. Principal among these were crabs of the genera *Menippe* (stone crabs) and *Calappa* (box crabs), which used their claws to peel back the shell from the aperture and expose the unprotected snail. Ecphoras with large healed shell breaks are common and attest to the high frequencies of unsuccessful crab attacks (Figure 3.3F,G). Ecphoras were also often victims of naticid predators, and conical, beveled drill holes are frequently encountered on juvenile specimens. Since ecphoras lived on sandy or muddy bottoms, which are also the ideal substrates for large moon snails, they were frequently vulnerable to this type of predation. In

A
B
C
D
E
F
G
H

the Salisbury Sea, the naticid genus *Euspira* was probably the principal molluscan predator on ecphoras, while in the Albemarle Sea, the main molluscan predators were moon snails of the genus *Neverita*.

ECPHORA BIOSTRATIGRAPHY

Within the Chesapeake Bay area, each ecphora species is restricted to a single Shattuck Zone, and the ecphoras, in general, are one of the best groups of stratigraphic index fossils in the Maryland Miocene. Once familiar with the shell morphologies of the 40 named Miocene ecphoras, even fragments of specimens can be used to determine the identity and age of any stratigraphic unit, anywhere in Maryland or Virginia. Interestingly enough, Ward (1992:121–129) recognized only four species and four subspecies of ecphoras for the entire Chesapeake Miocene. These include (using his taxonomy): *Ecphora tricostata* Martin, 1904 and *Ecphora tricostata pamlico* Wilson, 1987, Plum Point Member, Calvert Formation; *Ecphora meganae* Ward and Gilinsky, 1988, Calvert Beach Member, Calvert Formation, St. Leonard and Drumcliff Members, Choptank Formation; *Ecphora meganae williamsi* Ward and Gilinsky, 1988, Boston Cliffs Member, Choptank Formation; *Ecphora gardnerae germonae* Ward and Gilinsky, 1988, Little Cove Point Member, St. Mary's Formation; *Ecphora gardnerae* Wilson, 1987, Windmill Point Member, St. Mary's Formation; *Ecphora gardnerae whiteoakensis* Ward and Gilinsky, 1988, Claremont Manor Member, Eastover Formation; and *Ecphora kochi* Ward and Gilinski, 1988, Cobham Bay Member, Eastover Formation. All the other species shown in this book were considered by him to be synonyms of these eight taxa. This lumping of the other 32 taxa does not take into consideration the evolution and extinction patterns of the ecphora lineages discussed previously in this chapter and obfuscates any attempts at fine-tuned stratigraphy (see Petuch, 1988a, 1988b, 1989). All 40 ecphora species are illustrated in color in the downloadable Powerpoint on this title's product page on crcpress.com.

The ecphoras of the Maryland and Virginia Miocene are arranged here by stratigraphic unit. Only fossil-bearing units are listed.

Calvert Formation (Specimens illustrated in Chapter 4)

Fairhaven Member

Shattuck Zone 2

? *Siphoecphora aurora* (Wilson, 1987) (Only fragmentary specimens)

FIGURE 3.3 (Facing page) Paleoecological interactions of the Chesapeake Miocene ecphoras. **Drilling of prey:** A = *Mercenaria druidi gardnerae* drilled by *Ecphora whiteoakensis,* Claremont Manor Member, Eastover Formation, Urbanna Creek, Virginia; B = *Trisecphora patuxentia* drilled by another ecphora, Drumcliff Member, Choptank Formation, Patuxent River, Maryland; C = *Conradconfusus parilis* drilled by *Ecphora gardnerae,* Windmill Point Member, St. Mary's Formation, St. Mary's River, Maryland; D = *Mercenaria tetrica* drilled by *Ecphora gardnerae,* Windmill Point Member, St. Mary's Formation, St. Mary's River, Maryland; E = *Chesapecten ceccae* drilled by *Ecphora whiteoakensis,* Claremont Manor Member, Eastover Formation, Urbanna Creek, Virginia; H = *Dosinia thori* drilled by *Ecphora gardnerae,* Windmill Point Member, St. Mary's Formation, St. Mary's River, Maryland; **Failed predation attempts by crabs;** F = *Trisecphora bartoni* with two healed crab breaks, Calvert Beach Member, Calvert Formation, Calvert Cliffs, Maryland; G = *Ecphora gardnerae* with large crab break and deformed ribs, Windmill Point Member, St. Mary's Formation, St. Mary's River, Maryland.

Plum Point Member

Shattuck Zone 4

Trisecphora chamnessi (Petuch, 1989) (Figure 4.1B,C,I)

Shattuck Zone 10

Chesathais ecclesiasticus (Dall, 1915) (Figure 3.2D and 4.2F,G)
Chesathais whitfieldi Petuch, 1989 (Figure 4.4A)
Ecphora wardi Petuch, 1988 (Figure 4.3J,K)
Ecphorosycon pamlico (Wilson, 1987) (Figure 3.2A and 4.3A,B)
Trisecphora eccentrica (Petuch, 1989) (Figure 4.2A,B)
Trisecphora prunicola (Petuch, 1988) (Figure 3.2I and 4.3H,I)
Trisecphora schmidti (Petuch, 1989) (Figure 3.2G and 4.3F,G)
Trisecphora tricostata (Martin, 1904) (Figure 3.2E and 4.2F,G)

Shattuck Zone 12

Ecphora calvertensis Petuch, 1988 (Figure 3.1D and 4.6D)
Ecphorosycon kalyx (Petuch, 1988) (Figure 3.2B and 4.6M,N)
Trisecphora martini (Petuch, 1988) (Figure 3.2H and 4.6C)
Trisecphora tricostata subspecies (Figure 3.2F)

Calvert Beach Member

Shattuck Zone 14

Ecphora mattinglyi Petuch, 2004 (Figure 3.1F and 4.6A,B)
Trisecphora scientistensis (Petuch, 1992) (Figure 4.6K)

Shattuck Zone 16

Ecphora chesapeakensis Petuch, 1992 (Figure 3.1C and 4.6I,J)
Ecphora (Planecphora) turneri Petuch, 1992 (Figure 4.6E,F)
Trisecphora bartoni Petuch and Drolshagen, new species (Figure 3.2K and 4.6G,H)

Choptank Formation (Specimens illustrated in Chapter 5)

St. Leonard Member

Shattuck Zone 17

Chesathais donaldasheri Petuch, 1989 (Figure 5.1C)
Ecphora harasewychi Petuch, 1989 (Figure 3.1H and 5.1M)
Ecphora sandgatesensis Petuch, 1989 (Figure 5.1A,B)
Ecphora (Planecphora) vokesi Petuch, 1989 (Figure 5.1I,J)
Ecphorosycon lindajoyceae (Petuch, 1993) (Figure 5.1E,F)
Trisecphora smithae (Petuch, 1988) (Figure 5.1K,L)

Drumcliff Member

Shattuck Zone 18

Chesathais lindae Petuch, 1988 (Figure 5.2G,H)
Ecphora meganae Ward and Gilinsky, 1988 (Figure 3.1E and 5.2E,F)
Ecphora rikeri Petuch, 1988 (Figure 3.1G and 5.2A,B)
Ecphora (Planecphora) choptankensis Petuch, 1988 (Figure 3.1I and 5.2C,D)
Trisecphora patuxentia (Petuch, 1988) (Figure 3.2J and 5.2N,O)

Boston Cliffs Member

Shattuck Zone 19

Chesathais drumcliffensis Petuch, 1989 (Figure 3.2C and 5.4F)
Ecphora amyae Petuch and Drolshagen, new species (Figure 5.4J,K)

Ecphora williamsi Ward and Gilinsky, 1988 (Figure 3.1B and 5.4A,B)
Ecphora (Planecphora) delicata Petuch, 1989 (Figure 5.4C,D)
Trisecphora shattucki (Petuch, 1989) (Figure 5.4L)

St. Mary's Formation (Specimens illustrated in Chapter 6)
Conoy Member
Shattuck Zone 20
Ecphora conoyensis Petuch, 2004 (Figure 6.1G,H)
Little Cove Point Member
Shattuck Zone 21
Ecphora asheri Petuch, 1988 (Figure 6.1A,B)
Shattuck Zone 22
Ecphora germonae Ward and Gilinski, 1988 (Figure 6.2H,I)
Windmill Point Member
Shattuck Zone 24
Ecphora gardnerae Wilson, 1987 (Figure 3.1A and 6.3A,B)
Chancellor Point Sandstone
Ecphora gardnerae angusticostata Petuch, 1989 (Figure 6.8A,B)

Eastover Formation (Specimens illustrated in Chapter 7)
Claremont Manor Member
Ecphora whiteoakensis Ward and Gilinski, 1988 (Figure 7.1A,B)
Cobham Bay Member
Ecphora kochi Ward and Gilinski, 1988 (Figure 7.2A,B)

Two ecphora genera, *Ecphora* (sensu stricto) and *Trisecphora,* both produced a series of secondary species radiations that have biostratigraphic significance. As side branches off their main evolutionary lines, these separate offshoots (clades) are here referred to as "species complexes." Since the type species of the genus *Ecphora* is *E. quadricostata,* which has four flattened, "T"-shaped ribs, all similar Miocene species are included within this main lineage. For Maryland and Virginia this includes *Ecphora wardi, E. calvertensis, E. chesapeakensis, E. williamsi, E. conoyensis, E. asheri, E. germonae, E. gardnerae, E. whiteoakensis,* and *E. kochi.* A prominent side branch that broke off from this lineage during Shattuck Zone 14 time is the "*Ecphora meganae* Species Complex" (species with thinner ribs that are flattened on their edges and only slightly "T"-shaped in cross section). This complex, which died out at the end of Shattuck Zone 19 time, includes *Ecphora mattinglyi, E. sandgatesensis, E. megani,* and *E. amyae.* Another small offshoot of the main *Ecphora* line is the "*Ecphora rikeri* Species Complex," a short-lived group that is confined to the Choptank Formation (*Ecphora harasewychi* and *E. rikeri*).

The three-ribbed genus *Trisecphora,* similarly, evolved two separate offshoots off the main wide-ribbed line. As typified by the wide-ribbed ("T"-shaped in cross section) *Trisecphora tricostata* (the type species), the main lineage (the "*Trisecphora tricostata* Species Complex") contained only three species, including *T. tricostata, T. tricostata* subspecies, and *T. scientistensis.* By Shattuck Zone 14 time, this wide-ribbed *Trisecphora* species became extinct. One of its offshoots, the thin-ribbed species, continued on into Choptank time, becoming extinct in Shattuck Zone 19. Referred to here as the "*Trisecphora patuxentia* Species Complex," this group contained *Trisecphora prunicola, T. bartoni, T. smithae, T. patuxentia,* and *T. shattucki.* This was the longest-lived three-ribbed ecphora group, first appearing in Shattuck Zone 10 and disappearing in Shattuck Zone 19. The other *Trisecphora* offshoot was very short-lived, occurring only in Shattuck Zones 10 and 12. Referred to here as the "*Trisecphora schmidti* Species Complex," this clade contained only two species, *Trisecphora schmidti* and *T. martini.* Unlike all the other ecphoras, this species complex had upwardly curved ribs, giving them a cupped appearance. These and all the known ecphora species of the Chesapeake Miocene are discussed in context with their respective molluscan faunas in the following chapters.

4 Fossils of the Calvert Formation

Of the 16 Shattuck Zones and three members included within the Calvert Formation, only six contain fossils preserved well enough to be easily collected and identified. These include Shattuck Zone 2 of the Fairhaven Member, Shattuck Zones 4, 10, and 12 of the Plum Point Member, and Shattuck Zones 14 and 16 of the Calvert Beach Member. The molluscan faunas differ greatly between these members, reflecting the changing marine climates during the Burdigalian and Langhian Miocene. The Calvert Formation has been heavily leached by groundwater, and most of the fossils are soft and partially dissolved. Great care and patience must be taken when collecting Calvert mollusks, particularly aragonitic shells, because these often have the same consistency as the clay within which they are buried. Only durable calcitic shells such as oysters, scallops, and ecphoras are hard enough to be collectible in an intact state.

SHELLS OF THE FAIRHAVEN MEMBER

The oldest large molluscan fauna in the Chesapeake Miocene is found in Shattuck Zone 2 of the Fairhaven Member (Burdigalian Miocene). Accessible along the low cliffs south of Fairhaven, Ann Arundel County, and southward to north of Chesapeake Beach, Calvert County, Shattuck Zone 2 is generally a thin (less than 1 m) sand or sandstone layer containing an impoverished fauna of only around 15 species. The preservation is usually bad, with specimens being either moldic or fragmentary. A close look at the Shattuck Zone 2 fauna shows that it closely resembles that of Shattuck Zone 10 (Plum Point Member), indicating that they are not too far apart in time. Bivalves dominate this molluscan assemblage, some of which include *Lucinoma contracta* (Say, 1824), *Caryocorbula elevata* (Conrad, 1836), *Astarte cuneiformis* Conrad, 1840, *Astarte thomasi* Conrad, 1855, *Melosia* cf. *staminea* (Conrad, 1839), *Thracia conradi* Couthouy, 1939, *Panopea whitfieldi* Dall, 1898, and *Glossus mazlea* subspecies. These species, and the others listed at the end of this chapter, lived on open muddy bottom areas, probably on near-shore shallow banks. Extensive beds of scallops also lived on these shallow muddy bottoms, as evidenced by the large number of fragmentary specimens found in Shattuck Zone 2. At present, three species are known from this fauna, and these include *Chesapecten sayanus* (Dall, 1898), *Chesapecten coccymelus* subspecies, and *Pecten humphreysi* Conrad, 1842. Fragments of a three-ribbed ecphora, probably *Siphoecphora aurora* (Wilson, 1987) or an unnamed species, are also encountered in this fauna.

The uppermost unit of the Fairhaven Member, Shattuck Zone 3, contains very few species of mollusks, mostly partially dissolved specimens or molds of the bivalves *Caryocorbula elevata* (Conrad, 1836) and *Glossus mazlea* subspecies (see Chapter 2). Likewise, the first six Shattuck Zones of the Plum Point Member also have a very impoverished molluscan fauna, the result of the low sea level and cool marine climate at the end of the Burdigalian Age. Only Shattuck Zone 4 has a distinctive molluscan assemblage, often containing extensive beds of the gryphaeid oyster *Hyotissa percrassa* (Conrad, 1841) (Figure 4.1A; also see Chapters 1 and 2). The thick-ribbed ecphora *Trisecphora chamnessi* (Petuch, 1989)

A
B
C
D
E
F
G
H
I
J
K
L
M
N

(Figure 4.1B,C,I), ancestor of the Shattuck Zone 10 *Trisecphora eccentrica* (Petuch, 1989) (Figure 4.2A,B), has been collected in these oyster beds, but usually in a crushed and deformed state. Both species share the same type of thick, cordlike, ribs with distinct rounded edges. Shattuck Zones 5, 6, 7, 8, and 9 were all deposited in extremely shallow, near-shore muddy, brackish lagoons and only contain thin beds of the small bivalve *Caryocorbula elevata* (Conrad, 1836) (see Chapters 1 and 2).

SHELLS OF THE PLUM POINT MEMBER

The richest and best-preserved Calvert fauna is seen in Shattuck Zone 10 of the Plum Point Member. Near Plum Point, Calvert County, over 120 species of gastropods and bivalves can be collected (over 50 species illustrated here in Figure 4.1 through Figure 4.5), making this the second-richest molluscan fauna (after that of the Windmill Point Member of the St. Mary's Formation) in the entire Chesapeake Miocene. The Shattuck Zone 10 fauna contains a large component of tropical–subtropical genera, all of which do not range farther north than Cape Hatteras in the Recent. Their presence attests to the warm water conditions in the Salisbury Sea during the early Langhian Age. Principal among these are members of several classic tropical gastropod groups, some of which include the seilid *Seila clavulus* (H.C. Lea, 1843), the wentletraps *Stenorhytis pachypleura* (Conrad, 1841) (Figure 4.1L,M), *Cirsotrema calvertensis* (Martin, 1904) (Figure 4.1E), and *Amaea prunicola* (Martin, 1904) (Figure 4.1J), the melanellid *Niso lineata* (Conrad, 1841), the sundial shell *Architectonica trilineatum* (Conrad, 1841), the carrier shell *Xenophora lindae* Petuch and Drolshagen, new species (see Systematic Appendix) (Figure 4.4N,O), the fig shell *Ficus harrisi* (Martin, 1904) (Figure 4.5H), the olive shells *Oliva (Strephona) harrisi* Martin, 1904 (Figure 4.5D) and *Oliva (Strephona) simondsoni* Ward, 1998, the triton *Sassia centrosum* (Conrad, 1868), the pisaniaform buccinids *Celatoconus protractus* (Conrad, 1843) and *Metula calvertana* (Martin, 1904) (originally described as a *Siphonalia*), the marginellas *Dentimargo calvertensis* (Martin, 1904) and *Persicula conulus* (H.C. Lea, 1843), and an unnamed species of the auger shell *Myurellina* (similar to the Pliocene *Myurellina unilineata*). All these species have living relatives in the high tropical waters of the Caribbean, South Pacific, and Indian Ocean.

FIGURE 4.1 **(Facing page)** Index fossils for the Fairhaven and Plum Point Members of the Calvert Formation. A = *Hyotissa percrassa* (Conrad, 1841), length 103 mm, Shattuck Zone 4, Chesapeake Beach, Calvert County; B,C = *Trisecphora chamnessi* (Petuch, 1989), length 49 mm, Aurora, Beaufort County, North Carolina, Pungo River Formation (also found in Shattuck Zone 4); D = *Cyclocardia calvertensis* Petuch and Drolshagen, new species, (see Systematic Appendix), holotype, length 12 mm, Shattuck Zone 10, Plum Point, Calvert County; E = *Cirsotrema calvertensis* (Martin, 1904), length (incomplete) 28 mm; F = *Turrifulgur prunicola* Petuch, 1993, length 35 mm Shattuck Zone 10, Plum Point, Calvert County; G,H = *Turrifulgur marylandicus* Petuch, 1993, length (incomplete) 60 mm, Shattuck Zone 10, Plum Point, Calvert County; I = *Trisecphora chamnessi* (Petuch, 1989), length (crushed and deformed) 53 mm, Shattuck Zone 4, Chesapeake Beach, Calvert County; J = *Amaea prunicola* (Martin, 1904), length 20 mm, Aurora, Beaufort County, North Carolina, Pungo River Formation (also found in Shattuck Zone 10); K = *Busycotypus scalaspira* (Conrad, 1863), length (incomplete) 45 mm, Shattuck Zone 10, Plum Point, Calvert County; L,M = *Stenorhytis pachypleura* (Conrad, 1841), length 15 mm, Aurora, Beaufort County, North Carolina, Pungo River Formation (also found in Shattuck Zone 10); N = *Pecten humphreysi* Conrad, 1842, length 90 mm, Aurora, Beaufort County, North Carolina, Pungo River Formation (also found in Shattuck Zones 2 and 10).

A
B
C
D
E
F
G
H
I
J
K
L
M

The Shattuck Zone 10 bivalve fauna also contains a number of tropical–subtropical genera and species. Some of the more important of these are the ark shell *Barbatia marylandica* (Conrad, 1840), the bittersweet clam *Glycymeris parilis* (Conrad, 1843), the file shell *Lima papyria* Conrad, 1841, the pen shells *Atrina harrisi* Dall, 1898 and *Atrina piscatoria* Glenn, 1904, the leafy jewel box *Chama chipolana* subspecies, and the kitten's paw *Plicatula densata* Conrad, 1843. Besides these few tropical species, the bulk of the Shattuck Zone 10 bivalve fauna was dominated by typical Transmarian species such as the ark shell *Dallarca subrostrata* (Conrad, 1841), the venerids *Mercenaria blakei* Ward, 1992 (Figure 4.4B), *Melosia staminea* (Conrad, 1839) (Figure 4.2E), and *Dosinia blackwelderi* Ward, 1992 (Figure 4.3E), the lucinids *Phacoides foremani* (Conrad, 1841) (Figure 4.5E) and *Stewartia anodonta* (Say, 1824), the large corbulid *Bicorbula idonea* (Conrad, 1833) (Figure 4.5F), the astartids *Astarte cuneiformis* Conrad, 1840 (Figure 4.3M) and *Astarte calvertensis* Glenn, 1904, the bizarre, coiled glossid clams *Glossus marcoei* (Conrad, 1842) Figure 4.2J) and *Glossus mazlea* (Glenn, 1904) Figure 4.2L), the crassatellid *Marvacrassatella melina* (Conrad, 1832) (Figure 4.3C), the cockle shells *Chesacardium craticuloide* (Conrad, 1845)(Figure 4.5B), *Chesacardium calvertensium* (Glenn, 1904), and *Chesacardium leptopleurum* (Conrad, 1841), the scallop *Chesapecten coccymelus* (Dall, 1898), and the geoduck *Panopea whitfieldi* Dall, 1898. A distinct arctic component was also present in the Shattuck Zone 10 bivalve fauna, including the astartids *Astarte cuneiformis* Conrad, 1840 (Figure 4.3M), *Astarte vicinia* Say, 1824, *Astarte thomasii* Conrad, 1855, and *Astarte calvertensis* Glenn, 1904 and the cold-water crassatellids *Cyclocardia castrana* Glenn, 1904 and *Cyclocardia calvertensis* Petuch and Drolshagen, new species (see Systematic Appendix) (Figure 4.1D).

With the exception of the Arctic gastropod genera *Euspira* (Naticidae) and *Oenopota* (*O. cornelliana* (Martin, 1904), Turridae) the largest component of the Shattuck Zone 10 gastropod fauna comprised classic Transmarian Province endemic genera (see Chapter 1). Some of the more important of these include the muricid *Patuxentrophon plummerae* Petuch and Drolshagen, new species (see Systematic Appendix) (Figure 4.2H), the fasciolariids *Conradconfusus devexus* (Conrad, 1843) (Figure 4.5A) and *Conradconfusus migrans* (Conrad, 1843) (Figure 4.4K), the busycon whelks *Busycotypus scalaspira* (Conrad, 1863) (Figure 4.1K; *B. calvertensis* Petuch, 1988 may be a flat-spired variant of *B. scalaspira*), *Turrifulgur prunicola* Petuch, 1993 (Figure 4.1F), and *Turrifulgur marylandicus* Petuch, 1993 (Figure 4.1G,H), the volute *Scaphellopsis coronaspira* Petuch, 1988 (Figure 4.2K), the nutmeg shells *Marianarona engonata* (Conrad, 1841) (Figure 4.5J) and *Mariasveltia calvertensis* (Martin, 1904), and the turrids *Calverturris bellacrenata* (Conrad, 1841) (Figure 4.5I), *Hemipleurotoma protocommunis* (Martin, 1904) (Figure 4.5L), *Transmariaturris calvertensis* (Martin, 1904) (Figure 4.5G), *Mariaturricula calvertensis* (Martin, 1904) (Figure 4.5N), and *Sediliopsis calvertensis* (Martin, 1904).

FIGURE 4.2 (Facing page) Index fossils for Shattuck Zone 10, Plum Point Member of the Calvert Formation. A,B = *Trisecphora eccentrica* (Petuch, 1989), length 56 mm; C,D = *Trisecphora tricostata* (Martin, 1904), length 58 mm; E = *Melosia staminea* (Conrad, 1839), length 26 mm; F,G = *Chesathais ecclesiasticus* (Dall, 1915), length 50 mm; H = *Patuxentrophon plummerae* Petuch and Drolshagen, new species (see Systematic Appendix), holotype, length 12 mm; I = *Volutifusus caricelloides* Petuch, 1988, length 65 mm; J = *Glossus markoei* (Conrad, 1842), length 51 mm; K = *Scaphellopsis coronaspira* Petuch, 1988, length 35 mm; L = *Glossus mazlea* (Glenn, 1904), length 48 mm; M = *Chesapecten coccymelus* (Dall, 1898), length 53 mm. All specimens were collected in Shattuck Zone 10 at Plum Point, Calvert County, Maryland.

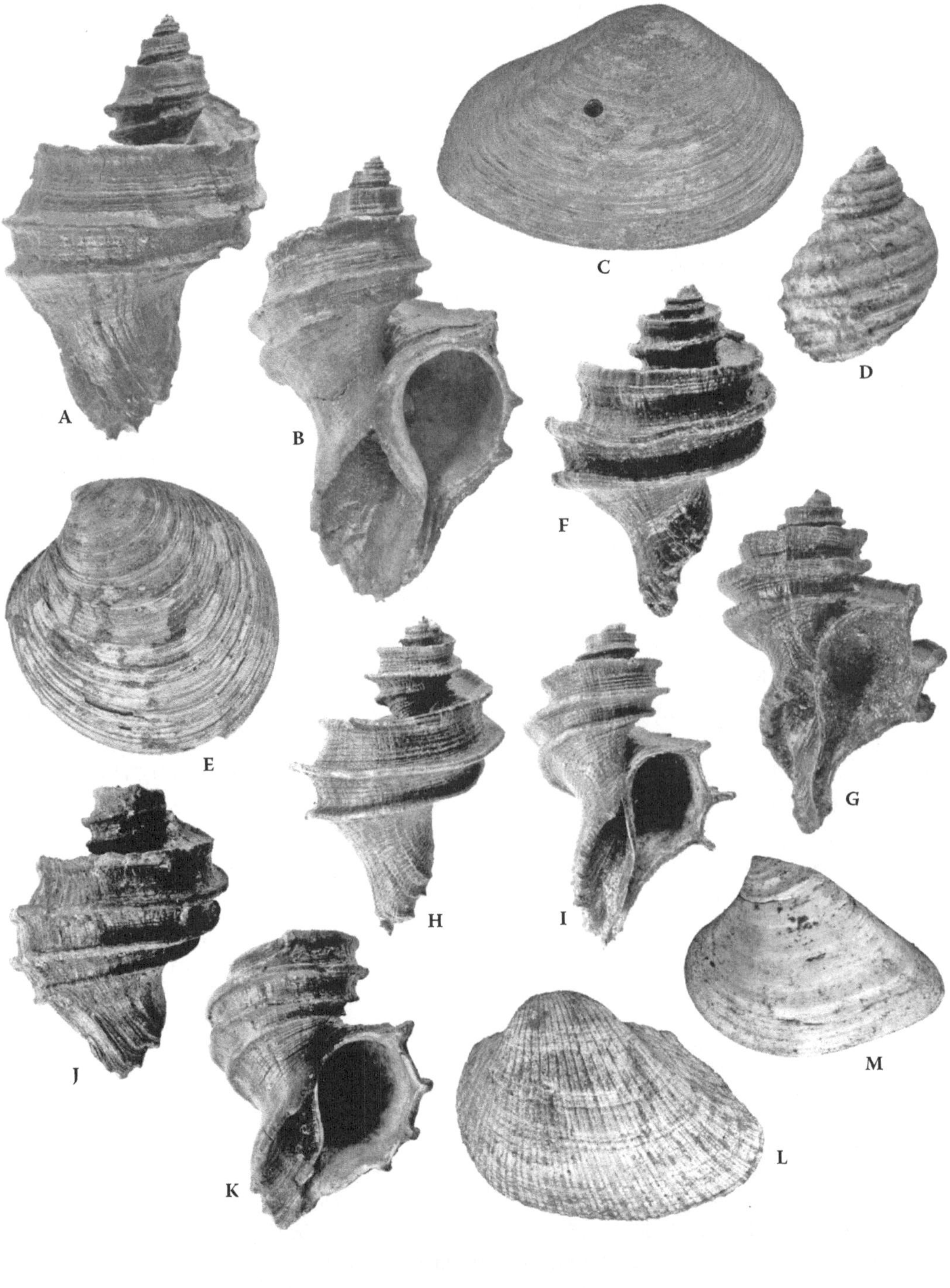
A
B
C
D
E
F
G
H
I
J
K
L
M

The largest species radiation of turritellid gastropods found in the Chesapeake Miocene is also a major component of the Shattuck Zone 10 molluscan fauna. This includes two species of the Transmarian endemic genus *Mariacolpus* Petuch, 1988, *M. aequistriata* (Conrad, 1863) (Figure 4.4K) and *M. prunicola* Petuch and Drolshagen, new species (see Systematic Appendix) (Figure 4.4L), two species of the subtropical-tropical genus *Torcula, T. exaltata* (Conrad, 1841) (Figure 4.4G) and *T. cumberlandiana* (Conrad, 1863) (Figure 4.4J), and a complex of the genus *Calvertitella* Petuch, 1988, including *Calvertitella indenta* (Conrad, 1841) (Figure 4.4I), *C. indenta calvertensis* (Oleksyshyn, 1959) (Figure 4.4M), *C. indenta chesapeakensis* (Oleksyshyn, 1959), and *C. indenta pagodula* Petuch, 1993. These last forms or subspecies may, upon closer study, prove to be full species. The Shattuck Zone 10 *Calvertitella* species are the last of a large group of distinctive turritellids that extends back to the Eocene (the *C. mortoni* complex) and Oligocene (the *C. tampae–C. fuerta* complex). *Calvertitella* becomes extinct at the end of the Calvert Formation, while the genera *Mariacolpus* and *Torcula* continue on and are found throughout the Choptank and St. Mary's Formations. The Chesapeake Miocene turritellids are often erroneously placed in the genus *Turritella* (i.e., Ward, 1992). That genus (the type of the family), however, is confined to the southwestern Pacific and Indian Oceans and is anatomically different from the American turritellids. There are no true *Turritella* (sensu stricto) species in the Americas, and only three genera are known from the Chesapeake Miocene.

As discussed in Chapter 3, the ecphora shells are a prominent part of the Shattuck Zone 10 fauna, with eight species having been collected to date. Two of the more unusual ecphoras include the bizarre, uncoiled, corkscrew-shaped *Trisecphora eccentrica* (Petuch, 1989) (Figure 4.2A,B; descendant of the lower Plum Point *T. chamnessi*) and *Trisecphora schmidti* (Petuch, 1989) (Figure 4.3F,G), which has cupped, upward-pointing ribs. These two relatively uncommon species differ from the common *Trisecphora tricostata* (Martin, 19040 (Figure 4.2C,D) in having consistently different shell shapes and rib structures. The commonly encountered *T. tricostata* is a large, inflated shell with a wide, open umbilicus and with wide, flattened ribs that are "T"-shaped in cross section (see Chapter 3). *Trisecphora eccentrica* is an elongated, uncoiled shell with thick, rounded cordlike ribs. Besides having distinctive upward-pointing ribs, *T. schmidti* differs from both *T. tricostata* and *T. eccentrica* in having a characteristic elongated siphonal canal and narrow umbilicus. The small, delicate, and elongated *T. prunicola* (Petuch, 1988) (Figure 4.3H,I) differs from these other three-ribbed species in having thin, bladelike ribs and coarse spiral threaded sculpture that covers the entire shell. Zone 10 contains the only ecphora fauna with four sympatric *Trisecphora* species.

Four other ecphoras are also found together with the *Trisecphora* species in Shattuck Zone 10. The largest of these is *Ecphorosycon pamlico* (Wilson, 1987) (Figure 4.3A,B), which occurs in, and was originally described from, the contemporaneous Pungo River Formation of northern North Carolina (Petuch, 1989; 2004:98–99). This large distinctive ecphora differs from the other smaller species in having thin, low, almost obsolete ribs

FIGURE 4.3 (Facing page) Index fossils for Shattuck Zone 10, Plum Point Member of the Calvert Formation. A,B = *Ecphorosycon pamlico* (Wilson, 1987), length 74 mm; C = *Marvacrassatella melina* (Conrad, 1832), length 62 mm; D = *Carinorbis dalli* (Whitfield, 1894), length 13 mm; E = *Dosinia blackwelderi* Ward, 1992, length 58 mm; F,G = *Trisecphora schmidti* (Petuch, 1989), length 30 mm; H,I = *Trisecphora prunicola* (Petuch, 1988), length 33 mm; J,K = *Ecphora wardi* Petuch, 1988, length 58 mm; L = *Dallarca subrostrata* (Conrad, 1841), length 28 mm; M = *Astarte cuneiformis* Conrad, 1840, length 27 mm. All specimens were collected in Shattuck Zone 10 at Plum Point, Calvert County, Maryland.

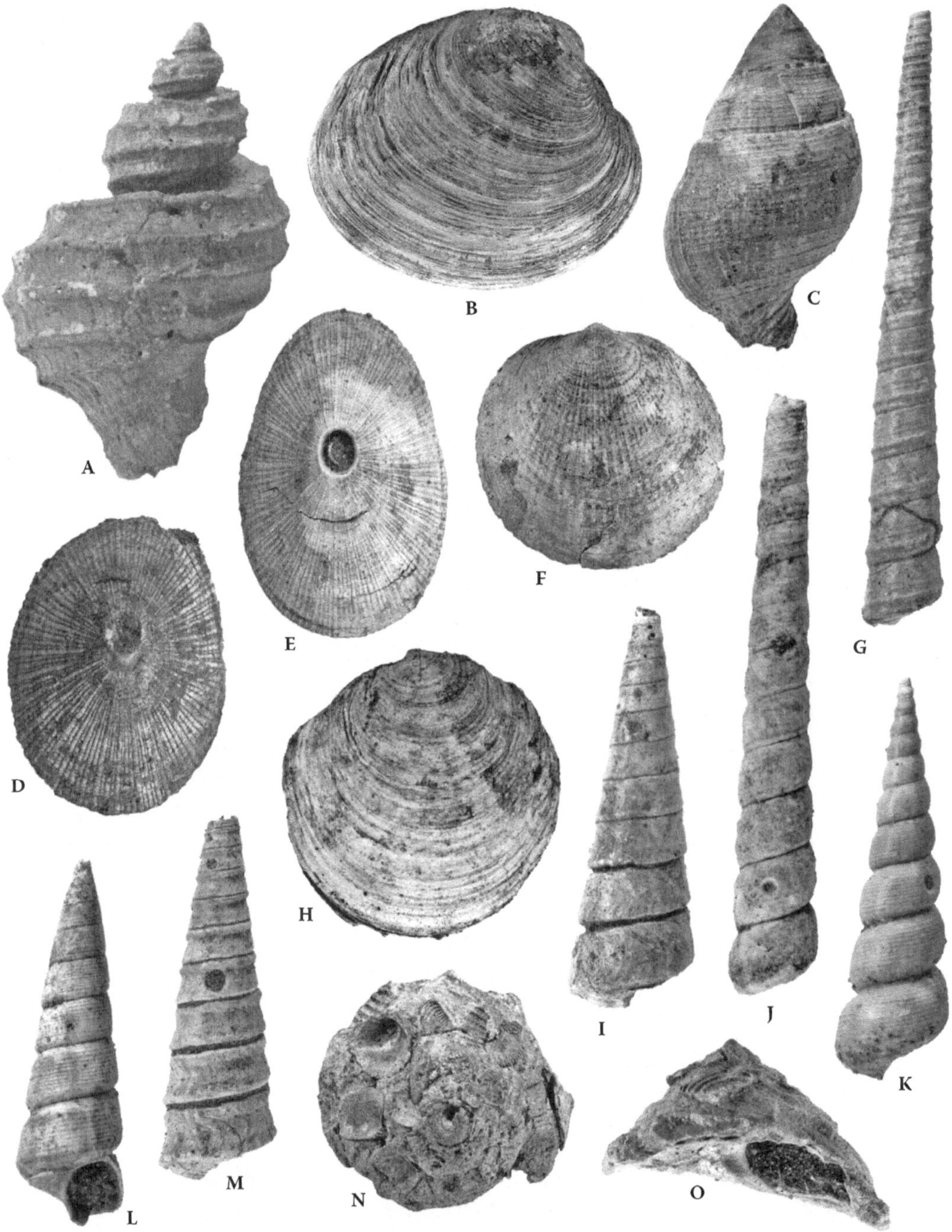
A
B
C
D
E
F
G
H
I
J
K
L
M
N
O

on the body and in having a shoulder rib that projects upward, producing a broad, sloping shoulder area. The first species of *Ecphora* (sensu stricto) in the Calvert Formation, with four equal-sized ribs, is *Ecphora wardi* Petuch, 1988 (Figure 4.3J,K). This uncommon species is the ancestor of all the four-ribbed species complexes found in the Chesapeake Miocene and the later Pliocene formations. The rarest Calvert ecphoras belong to the genus *Chesathais,* and include *Chesathais ecclesiasticus* (Dall, 1915) (Figure 4.2F,G; named for Church Hill, Maryland) and *C. whitfieldi* Petuch, 1989 (Figure 4.4A; originally named from a New Jersey specimen). These odd, multiribbed shells have large cords inside the aperture, a feature that is seldom seen in other ecphora groups.

At the end of Shattuck Zone 10 time, in the late Langhian Age, the warm marine climate began to deteriorate, ushering in a time of colder water conditions. This led to a large extinction of tropical–subtropical forms, referred to as the "Transmarian Extinction" (Petuch, 1993; 2004:261–262; see Chapter 1 of this book). Disappearing permanently from the Maryland fossil record at this time were the gastropod genera *Cerithiopsis, Seila, Carinorbis* (Figure 4.3D), *Cirsotrema, Amaea, Niso, Architectonica, Erato, Xenophora, Ficus, Sassia, Murexiella, Oliva, Persicula, Myurellina,* and *Glyphostoma,* the bivalve genera *Barbatia, Atrina, Plicatula, Chama, Lima,* and *Divalinga,* and the nautiloid *Aturia* (known only from septal fragments; see Martin, 1904, and the illustration on the dedication page of this book). Lowered sea levels due to this global cooling event led to the deposition of thick clay layers in much of the Salisbury Sea, probably due to the delta systems along the shorelines of Maryland, Delaware, and New Jersey (Millville Delta; see Chapter 1). The uppermost units of the Plum Point Member, Shattuck Zones 12 and 13, were deposited in these clay-rich deltaic conditions and were heavily leached by infiltrating groundwater. As a result, all the aragonitic shells of Shattuck Zone 12 have been completely or partially dissolved away, leaving only the harder, calcitic shell fossils. Often in a fragmentary or crushed state, only a few ecphora species are known from Shattuck Zone 12. These include the distinctive wide-ribbed *Ecphora calvertensis* Petuch, 1988 (the descendant of *E. wardi* and the first of the four-ribbed ecphoras with wide, "T"-shaped ribs) (Figure 4.6D), the large, slope-spired *Ecphorosycon kalyx* (Petuch, 1988) (Figure 4.6M,N; the descendant of *E. pamlico*), an undescribed subspecies of *Trisecphora tricostata* (smaller and more cylindrical, with very wide ribs; Figure 3.2F), and *Trisecphora martini* (Petuch, 1988) (with cupped, upward-turning ribs and the descendant of *T. schmidti;* Figure 4.6C). Broken fragments of an undescribed *Chesathais* species have also been found in Shattuck Zone 12, as have fragments of a thin-ribbed *Trisecphora* that appears to be the descendant of *T. prunicola*.

FIGURE 4.4 **(Facing page)** Index fossils for Shattuck Zone 10, Plum Point Member of the Calvert Formation. A = *Chesathais whitfieldi* Petuch, 1989, length 57 mm; B = *Mercenaria blakei* Ward, 1992, length 74 mm; C = *Ptychosalpinx lienosa* (Conrad, 1843), length 50 mm; D = *Glyphis griscombi* (Conrad, 1834), length 34 mm; E = *Glyphis marylandica* (Conrad, 1841), length 37 mm; F = *Glycymeris parilis* (Conrad, 1843), length 42 mm; G = *Torcula exaltata* (Conrad, 1841), length 93 mm; H = *Stewartia anodonta* (Say, 1824), length 44 mm; I = *Calvertitella indenta* (Conrad, 1841), length 41 mm; J = *Torcula cumberlandiana* (Conrad, 1863), length 90 mm; K = *Mariacolpus aequistriata* (Conrad, 1863), length 31 mm; L = *Mariacolpus prunicola* Petuch and Drolshagen, new species (see Systematic Appendix), holotype, length 27 mm; M = *Calvertitella indenta calvertensis* (Oleksyshyn, 1959), length 40 mm; N,O = *Xenophora lindae* Petuch and Drolshagen, new species (see Systematic Appendix), holotype, length 31 mm. All specimens were collected in Shattuck Zone 10 at Plum Point, Calvert County, Maryland.

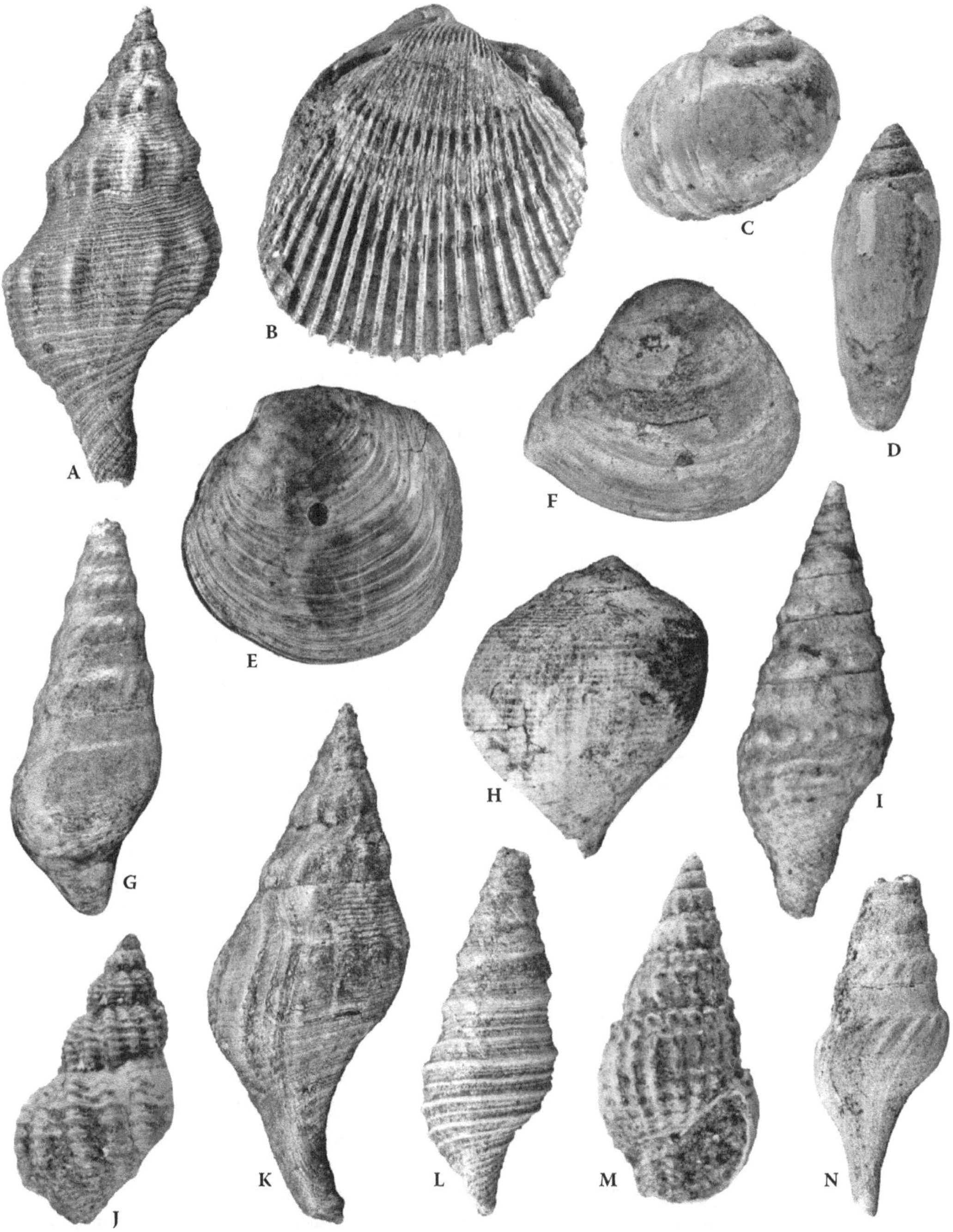
A
B
C
D
E
F
G
H
I
J
K
L
M
N

SHELLS OF THE CALVERT BEACH MEMBER

The last member of the Calvert Formation, the Calvert Beach Member, contains an impoverished and poorly preserved fauna similar to that of Shattuck Zone 12. The lower bed of the Calvert Beach Member, Shattuck Zone 14, contains only two described ecphora species, although fragments of at least three other undescribed species are frequently encountered. The named species include the large *Ecphora mattinglyi* Petuch, 2004 (Figure 4.6A,B; first species of the *Ecphora meganae* Species Complex and the ancestor of the Choptank *E. meganae*) and *Trisecphora scientistensis* (Petuch, 1992) (Figure 4.6K; descendant of *T. tricostata*). Fragments of undescribed species of *Ecphorosycon, Chesathais,* and a thin-ribbed *Trisecphora* are also known from this Shattuck Zone. During Shattuck Zone 14 time, the cupped-ribbed *Trisecphora schmidti* Species Complex became extinct, as that rib morphology is not seen again in the Maryland fossil record.

The upper bed of the Calvert Beach Member, Shattuck Zone 16, like Shattuck Zones 12 and 14, contains an impoverished molluscan fauna composed primarily of badly preserved ecphoras. Three species have been described from Shattuck Zone 16, including the wide-ribbed *Ecphora chesapeakensis* Petuch, 1992 (Figure 4.6I,J; descendant of *E. calvertensis* and the ancestor of the Choptank *E. williamsi*), the vase-shaped *Trisecphora bartoni* Petuch and Drolshagen, new species (see Systematic Appendix) (Figure 4.6G,H), and the first species of the thin-ribbed species complex, *Ecphora (Planecphora) turneri* Petuch, 1992 (Figure 4.6E,F). This last species is ancestral to the thin-ribbed *Ecphora (Planecphora) choptankensis* Species Complex of the Choptank Formation. The wide-ribbed *Trisecphora* species, characterized by *T. tricostata* and *T. scientistensis,* also became extinct at this time, leaving only one lineage of three-ribbed species (the *Trisecphora patuxentia* Species Complex of the Choptank Formation). Fragments of undescribed species of *Ecphorosycon* and *Chesathais* species are also found within Shattuck Zone 16, but no complete specimens have been found to date. Besides gastropods, a rare, poorly preserved bivalve species is occasionally encountered in Shattuck Zone 16, the venerid *Lirophora parkeria* (Glenn, 1904) (Figure 4.6L).

SYSTEMATIC LIST OF CALVERT MOLLUSKS

The following is a list of the more important index fossils of the Calvert Formation, arranged systematically and by Shattuck Zone. For illustrations and descriptions of micromollusks and tiny species, see Martin (1904) and Glenn (1904). Only Shattuck Zones that contain fossils are listed.

FIGURE 4.5 (Facing page) Index fossils for Shattuck Zone 10, Plum Point Member of the Calvert Formation. A = *Conradconfusus devexus* (Conrad, 1843), length 62 mm; B = *Chesacardium craticuloide* (Conrad, 1845), length 64 mm; C = *Euspirella tuomeyi* (Whitfield, 1894), length 14 mm; D = *Oliva (Strephona) harrisi* Martin, 1904, length 24 mm; E = *Phacoides foremani* (Conrad, 1841), length 43 mm; F = *Bicorbula idonea* (Conrad, 1833), length 27 mm; G = *Transmariaturris calvertensis* (Martin, 1904), length 40 mm; H = *Ficus harrisi* (Martin, 1904), length 29 mm; I = *Calverturris bellacrenata* (Conrad, 1841), length 28 mm; J = *Marianarona engonata* (Conrad, 1841), length 20 mm; K = *Conradconfusus migrans* (Conrad, 1843), length 65 mm; L = *Hemipleurotoma protocommunis* (Martin, 1904), length 25 mm; M = *Ilyanassa trivittatoides* (Whitfield, 1894), length 12 mm; N = *Mariaturricula marylandica* (Conrad, 1841), length 28 mm. All specimens were collected in Shattuck Zone 10 at Plum Point, Calvert County, Maryland.

A
B
C
D
E
F
G
H
I
J
K
L
M
N

CALVERT FORMATION

Fairhaven Member

Shattuck Zone 2

Gastropoda

Muricidae

?*Siphoecphora aurora* (Wilson, 1987) (known from fragments only)

Bivalvia

Pectinidae

Chesapecten sayanus (Dall, 1898)

Chesapecten coccymelus subspecies

Pecten humphreysi (Conrad, 1842) (Figure 4.1N)

Lucinidae

Lucinoma contracta (Say, 1824)

Corbulidae

Caryocorbula elevata (Conrad, 1836)

Caryocorbula whitfieldi (Dall, 1898)

Veneridae

Melosia cf. *staminea* (Conrad, 1839)

Thraciidae

Thracia conradi Couthouy, 1839

Crassatellidae

Cyclocardium calvertensis Petuch and Drolshagen, new species (Figure 4.1D)

Astartidae

Astarte cuneiformis Conrad, 1840 (Figure 4.3M)

Astarte thomasi Conrad, 1855

Glossidae

Glossus mazlea subspecies

Hiatellidae

Panopea americana Conrad, 1838

Panopea whitfieldi Dall, 1898

Shattuck Zone 3

Bivalvia

FIGURE 4.6 (Facing page) Index fossils for the Plum Point and Calvert Beach Members of the Calvert Formation. A,B = *Ecphora mattinglyi* Petuch, 2004, length 88 mm, Shattuck Zone 14, south of Chesapeake Beach, Calvert County; C = *Trisecphora martini* (Petuch, 1988), length 25 mm, Shattuck Zone 12, north of Scientist's Cliffs, Calvert County; D = *Ecphora calvertensis* Petuch, 1988, length 43 mm, Shattuck Zone 12, north of Scientist's Cliffs, Calvert County; E,F = *Ecphora (Planecphora) turneri* Petuch, 1992, length 40 mm, Shattuck Zone 16, north of Governor Run, Calvert County; G,H = *Trisecphora bartoni* Petuch and Drolshagen, new species (see Systematic Appendix), holotype, length 34 mm, Shattuck Zone 16, north of Governor Run, Calvert County; I,J = *Ecphora chesapeakensis* Petuch, 1992, length 41 mm, Shattuck Zone 16, north of Governor Run, Calvert County; K = *Trisecphora scientistensis* (Petuch, 1992), Shattuck Zone 14, north of Scientist's Cliffs, Calvert County; L = *Lirophora parkeria* (Glenn, 1904), length 26 mm, Shattuck Zone 16, north of Governor Run, Calvert County; M,N = *Ecphorosycon kalyx* (Petuch, 19880, length 68 mm, Shattuck Zone 12, north of Scientist's Cliffs, Calvert County.

Corbulidae
Caryocorbula elevata (Conrad, 1836)
Glossidae
Glossus mazlea subspecies

Plum Point Member
Shattuck Zone 4
Gastropoda
Muricidae
Trisecphora chamnessi (Petuch, 1989) (Figure 4.1B,C,I)
Bivalvia
Gryphaeidae
Hyotissa percrassa (Conrad, 1841) (Figure 4.1A)
Shattuck Zones 5–9
Bivalvia
Corbulidae
Caryocorbula elevata (Conrad, 1836)
Shattuck Zone 10
Gastropoda
Fissurellidae
Glyphis griscombi (Conrad, 1834) (Figure 4.4D)
Glyphis marylandica (Conrad, 1841) (Figure 4.4E)
Emarginula marylandica Martin, 1904
Trochidae
Calliostoma calvertanum Martin, 1904
Calliostoma marylandicum Martin, 1904
Calliostoma peralveatum (Conrad, 1841)
Turritellidae
Calvertitella indenta (Conrad, 1841) (Figure 4.4I)
Calvertitella indenta calvertensis (Oleksyshyn, 1959) (Figure 4.4M)
Calvertitella indenta chesapeakensis (Oleksyshyn, 1959)
Calvertitella indenta pagodula Petuch, 1993
Mariacolpus aequistriata (Conrad, 1863) (Figure 4.4K)
Mariacolpus prunicola Petuch and Drolshagen, new species (Figure 4.4L)
Torcula cumberlandiana (Conrad, 1863) (Figure 4.4J)
Torcula exaltata (Conrad, 1841) (Figure 4.4G)
Cerithiopsidae
Cerithiopsis calvertensis Martin, 1904
Seilidae
Seila clavulus (H.C. Lea, 1843)
Fossaridae
Carinorbis dalli (Whitfield, 1894) (Figure 4.3D)
Crepidulidae
Crepidula lamina H.C. Lea, 1843
Calyptraea gyrinum (Conrad, 1843)
Crucibulum cf. *costatum* (Say, 1820)

Epitoniidae
- *Amaea prunicola* (Martin, 1904) (Figure 4.1J)
- *Amaea reticulata* (Martin, 1904)
- *Cirsotrema calvertensis* (Martin, 1904) (Figure 4.1E)
- *Stenorhytis pachypleura* (Martin, 1904) (Figure 4.1L,M)

Melanellidae
- *Niso lineata* (Conrad, 1841)

Architectonicidae
- *Architectonica trilineatum* (Conrad, 1841)

Eratoidae
- *Erato perexigua* (Conrad, 1841)

Xenophoridae
- *Xenophora lindae* Petuch and Drolshagen, new species (Figure 4.4N,O)

Naticidae
- *Euspira* species (*E. heros* complex)
- *Euspirella tuomeyi* (Whitfield, 1894) (Figure 4.5C)
- *Sinum* cf. *fragilis* (Conrad, 1830)

Ficidae
- *Ficus harrisi* (Martin, 1904) (Figure 4.5H)

Personiidae
- *Sassia centrosum* (Conrad, 1868)

Muricidae
- *Chesathais ecclesiasticus* (Dall, 1915) (Figure 4.2F,G)
- *Chesathais whitfieldi* Petuch, 1989 (Figure 4.4A)
- *Ecphora wardi* Petuch, 1988 (Figure 4.3J,K)
- *Ecphorosycon pamlico* (Wilson, 1987) (Figure 4.3A,B)
- *Trisecphora eccentrica* (Petuch, 1989) (Figure 4.2A,B)
- *Trisecphora prunicola* (Petuch, 1988) (Figure 4.3H,I)
- *Trisecphora schmidti* (Petuch, 1989) (Figure 4.3F,G)
- *Trisecphora tricostata* (Martin, 1904) (Figure 4.2C,D)
- *Laevityphis* species
- *Murexiella shilohensis* (Heilprin, 1887)
- *Patuxentrophon plummerae* Petuch and Drolshagen, new species (Figure 4.2H)

Fasciolariidae
- *Conradconfusus devexus* (Conrad, 1843) (Figure 4.5A)
- *Conradconfusus migrans* (Conrad, 1843) (Figure 4.5K)

Buccinidae
- *Celatoconus protractus* (Conrad, 1843)
- *Metula calvertana* (Martin, 1904)
- *Ptychosalpinx lienosa* (Conrad, 1843) (Figure 4.4C)
- *Ptychosalpinx* species

Busyconidae
- *Busycotypus scalaspira* (Conrad, 1863) (Figure 4.1K)
- *Busycotypus scalaspira* form *calvertensis* Petuch, 1988
- *Sycopsis* species
- *Turrifulgur marylandicus* Petuch, 1993 (Figure 4.1G,H)

Turrifulgur prunicola Petuch, 1993 (Figure 4.1F)
Nassariidae
Ilyanassa calvertensis (Martin, 1904)
Ilyanassa greensboroensis (Martin, 1904)
Ilyanassa trivittatoides (Whitfield, 1894) (Figure 4.5M)
Columbellidae
Mitrella calvertensis (Martin, 1904)
Volutidae
Scaphellopsis coronaspira Petuch, 1988 (Figure 4.2K)
Volutifusus caricelloides Petuch, 1988 (Figure 4.2I)
Olividae
Oliva (Strephona) harrisi Martin, 1904 (Figure 4.5D)
Oliva (Strephona) simondsoni Ward, 1998
Marginellidae
Dentimargo calvertensis (Martin, 1904)
Persicula conulus (H.C. Lea, 1843)
Cancellariidae
Marianarona engonata (Conrad, 1841) (Figure 4.5J)
Marianarona prunicola (Martin, 1904)
Marianarona reticulatoides (Martin, 1904)
Mariasveltia calvertensis (Martin, 1904)
Turridae
Calverturris bellacrenata (Conrad, 1841) (Figure 4.5I)
Chesaclava pseudoeburnea (Heilprin, 1887)
Cymatosyrinx species
Hemipleurotoma protocommunis (Martin, 1904) (Figure 4.5L)
Mariaturricula calvertensis (Martin, 1904)
Mariaturricula marylandica (Conrad, 1841) (Figure 4.5N)
Oenopota cornelliana (Martin, 1904)
Sediliopsis calvertensis (Martin, 1904)
Transmariaturris calvertensis (Martin, 1904) (Figure 4.5G)
Clathurellidae
Glyphostoma obtusa (Martin, 1904)
Terebridae
Myurellina species
Strioterebrum calvertensis (Martin, 1904)
Strioterebrum dalli (Martin, 1904)
Strioterebrum whitfieldi (Martin, 1904)
Acteonidae
Acteon calvertensis Martin, 1904
Bivalvia
Arcidae
Barbatia marylandica (Conrad, 1840)
Dallarca subrostrata (Conrad, 1841) (Figure 4.3L)
Glycymeridae
Glycymeris parilis (Conrad, 1843) (Figure 4.4F)

Pectinidae
- *Chesapecten coccymelus* (Dall, 1898) (Figure 4.2M)
- *Eburneopecten cerinus* (Conrad, 1869)
- *Pecten humphreysi* Conrad, 1842 (Figure 4.1N)

Limidae
- *Lima papyria* Conrad, 1841

Pinnidae
- *Atrina harrisi* Dall, 1898
- *Atrina piscatoria* Glenn, 1904

Gryphaeidae
- *Hyotissa percrassa* (Conrad, 1841) (Figure 4.1A)

Plicatulidae
- *Plicatula densata* Conrad, 1843

Mytilidae
- *Modiolaria curta* Glenn, 1904
- *Perna* cf. *conradiana* (d'Orbigny, 1852)

Chamidae
- *Chama* species (*C. chipolanus* complex)

Lucinidae
- *Divalinga* species
- *Lucinoma contracta* (Say, 1824)
- *Parvilucina prunus* (Dall, 1903)
- *Phacoides foremani* (Conrad, 1841) (Figure 4.5E)
- *Phacoides trisulcatus* (Conrad, 1841)
- *Stewartia anodonta* (Say, 1824) (Figure 4.4H)

Corbulidae
- *Bicorbula idonea* (Conrad, 1833) (Figure 4.5F)
- *Caryocorbula elevata* (Conrad, 1838)
- *Caryocorbula whitfieldi* (Dall, 1898)

Veneridae
- *Dosinia blackwelderi* Ward, 1992 (Figure 4.3E)
- *Melosia staminea* (Conrad, 1839) (Figure 4.2E)
- *Mercenaria blakei* Ward, 1992 (Figure 4.4B)

Mactridae
- *Hemimactra chesapeakensis* (Glenn, 1904)

Tellinidae
- *Tellina declivis* Conrad, 1834
- *Tellina producta* Conrad, 1840

Thraciidae
- *Thracia conradi* Couthouy, 1839

Cardiidae
- *Chesacardium calvertensium* (Glenn, 1904)
- *Chesacardium craticuloide* (Conrad, 1845) (Figure 4.5B)
- *Chesacardium leptopleurum* (Conrad, 1841)

Crassatellidae
- *Cyclocardia calvertensis* Petuch and Drolshagen, new species (Figure 4.1D)

Cyclocardia castrana Glenn, 1904
Marvacrassatella melina (Conrad, 1832) (Figure 4.3C)
Astartidae
Astarte calvertensis Glenn, 1904
Astarte cuneiformis Conrad, 1840 (Figure 4.3M)
Astarte thomasi Conrad, 1855)
Astarte vicinia Say, 1824
Glossidae
Glossus markoei (Conrad, 1842) (Figure 4.2J)
Glossus mazlea (Glenn, 1904) (Figure 4.2L)
Hiatellidae
Panopea americana Conrad, 1838
Panopea whitfieldi Dall, 1898
Cephalopoda
Nautiloidea
Nautilidae
Aturia species (known only from septal fragments) (drawing on dedication page)

Shattuck Zone 12
Gastropoda
Muricidae
Ecphora calvertensis Petuch, 1988 (Figure 4.6D)
Ecphorosycon kalyx (Petuch, 1988) (Figure 4.6M,N)
Trisecphora martini (Petuch, 1988) (Figure 4.6C)
Trisecphora tricostata subspecies

Calvert Beach Member

Shattuck Zone 14
Gastropoda
Muricidae
Ecphora mattinglyi Petuch, 2004 (Figure 4.6A,B)
Trisecphora scientistensis (Petuch, 1992) (Figure 4.6K)
Bivalvia
Glossidae
Glossus mazlea subspecies

Shattuck Zone 16
Gastropoda
Muricidae
Ecphora chesapeakensis Petuch, 1992 (Figure 4.6I,J)
Ecphora (Planecphora) turneri Petuch, 1992 (Figure 4.6E,F)
Trisecphora bartoni Petuch and Drolshagen, new species (Figure 4.6G,H)
Bivalvia
Veneridae
Lirophora parkeria (Glenn, 1904) (Figure 4.6L)

(see Petuch, 1988c, 1992, 1993, 1997, and 2004)

5 Fossils of the Choptank Formation

The Choptank Formation differs from the Calvert Formation in that all its members contain well-preserved fossil mollusks. Having been deposited during the Mid-Miocene Cold Time, in a period of rapidly fluctuating sea levels and lowered water temperatures (Petuch, 2004:101, 261), the Choptank molluscan fauna is relatively impoverished. The three members of the Choptank Formation are each represented by a single Shattuck Zone, with Shattuck Zone 17 being the St. Leonard Member, Shattuck Zone 18 being the Drumcliff Member, and Shattuck Zone 19 being the Boston Cliffs Member (Ward, 1992). Of these members, the St. Leonard Member contains the most impoverished fauna, while the Drumcliff Member contains the richest fauna with the highest diversity.

SHELLS OF THE ST. LEONARD MEMBER

As the most leached unit of the Choptank, the St. Leonard Member (Shattuck Zone 17) contains mostly calcitic fossils. Of these, the ecphoras are the most abundant and prominent. The ecphora fauna of Zone 17 is also the largest of the Choptank members, with at least six species in five genera. Of these, the most frequently encountered and abundant is the thin-ribbed *Ecphora (Planecphora) vokesi* Petuch, 1989 (Figure 5.1I,J) (erroneously listed by Martin (1904) as "*Ecphora quadricostata umbilicata* Wagner," in reality a species from the Pliocene Yorktown Formation). This characteristic Shattuck Zone 17 species is the descendant of *Ecphora (Planecphora) turneri* of Shattuck Zone 16 (Calvert Beach Member, Calvert Formation) and differs in being a larger shell with lower, less-projecting ribs, and in having a much wider, flaring umbilicus. Another common Shattuck Zone 17 ecphora is the three-ribbed *Trisecphora smithae* (Petuch, 1988) (Figure 5.1K,L). This large species is the descendant of the Shattuck Zone 16 *Trisecphora bartoni* and differs in being a much larger, inflated shell with a rounded, not vase-shaped, profile.

The other locally common St. Leonard ecphora is *Ecphora sandgatesensis* Petuch, 1989 (Figure 5.1A,B), a small species descended from the Shattuck Zone 16 *Ecphora mattinglyi.* Although having the same type of rib structure as *E. mattinglyi* (ribs flattened along the edges and slightly "T"-shaped in cross section and being ornamented with a single incised groove), *E. sandgatesensis* differs from its Calvert ancestor in being a smaller, more cylindrical shell with the posteriormost (uppermost) rib projecting upward (posteriorly), producing deeply channeled, cupped spire whorls. This characteristic projecting shoulder rib is often ornamented with coarse spiral threads. *Ecphora sandgatesensis* is most often encountered near the Sandgates and Queen's Cliff areas, St. Mary's County, along the Patuxent River.

Besides the three commonly collected species, three other rare ecphoras are also found in Shattuck Zone 17. The rarest of these is *Ecphorosycon lindajoyceae* (Petuch, 1993) (Figure 5.1E,F), the largest ecphora found in the St. Leonard Member. The descendant of the Shattuck Zone 12 *Ecphorosycon kalyx, E. lindajoyceae* differs in being a larger, much

A
B
C
D
E
F
G
H
I
J
K
L
M
N

more inflated and rounded shell with a lower spire. The spiral threaded sculpture of *E. lindajoyceae* is also much finer and less prominent than that of *E. kalyx,* giving the shell surface a silky appearance. *Ecphorosycon lindajoyceae* is the last species of its genus, which died out at the end of St. Leonard time. To date, *E. lindajoyceae* has only been collected from the Sandgates–Queen's Cliff area along the Patuxent River and is known from fewer than a dozen specimens.

Two other rare St. Leonard species are also known primarily from the Sandgates area, and include *Chesathais donaldasheri* Petuch, 1989 (Figure 5.1C) and *Ecphora harasewychi* Petuch, 1989 (Figure 5.1M). The heavily sculptured *Chesathais donaldasheri* differs from its Calvert Formation ancestor, *Chesathais ecclesiasticus,* in being a smaller, vase-shaped shell with a distinctive, upward-pointing, flattened shoulder rib. This extended shoulder rib is characteristically ornamented with multiple strong, smaller ribs and fine cords. The rhomboid-shaped *E. harasewychi* represents the first appearance of a new ecphora group, the *Ecphora rikeri* Species Complex. Members of this offshoot group differ from other *Ecphora* species in having thinner, more rounded ribs that adhere to the body whorl and do not project outward.

Besides these six ecphora species, other muricid groups are conspicuously absent from Shattuck Zone 17. Only one exception has ever been discovered, the rare muricine *Stephanosalpinx candelabra* Petuch, 1988 (Figure 5.1G,H). This small species, with its large shoulder spikes and labial tooth is morphologically closest to the tropical genus *Panamurex* and may represent an endemic Transmarian Province offshoot. This unusual muricid is presently known from only a single broken specimen collected below Matoaka Cottages, St. Leonard, Calvert County, making it one of the rarest Maryland fossils. The immature specimen of a "*Trophon* species" listed by Martin (1904:204) from the Choptank Formation at Greensboro (specimen lost) may be this species. Besides ecphoras, the only other common gastropod in Shattuck Zone 17 is the small scaphelline volute *Volutifusus conradianus* (Martin, 1904) (Figure 5.1N), which is almost always poorly preserved and in a soft, fragile condition. A few other gastropods are known from the St. Leonard Member, the most frequently encountered being the trochid *Calliostoma* species (*C. philanthropus* complex), the turritellid *Mariacolpus octonaria* subspecies, the naticid moon snail *Euspira* species (*E. heros* complex), and the nassariid mud snail *Ilyanassa gubernatoria* (Martin, 1904) (named for Governor Run).

The bivalves of the St. Leonard Member, likewise, show a low diversity, with only a few large species being present. Principal among these is the venerid *Mercenaria cuneata* Conrad, 1867 (Figure 5.1D), which occasionally occurs in large beds. Other species include the ark shell *Dallarca* cf. *staminea* (Say, 1832), the large scallop *Chesapecten nephrens*

FIGURE 5.1 **(Facing page)** Index fossils for the St. Leonard Member of the Choptank Formation (Shattuck Zone 17). A,B = *Ecphora sandgatesensis* Petuch, 1989, length 40 mm, Sandgates, St. Mary's County, Patuxent River; C = *Chesathais donaldasheri* Petuch, 1989, length 23 mm, Sandgates, St. Mary's County, Patuxent River; D = *Mercenaria cuneata* Conrad, 1867, length 98 mm, Matoaka Cottages, St. Leonard, Calvert County; E,F = *Ecphorosycon lindajoyceae* (Petuch, 1993), length 82 mm, Sandgates, St. Mary's County, Patuxent River; G,H = *Stephanosalpinx candelabra* Petuch, 1988, length (incomplete) 18 mm, Matoaka Cottages, St. Leonard, Calvert County; I,J = *Ecphora (Planecphora) vokesi* Petuch, 1988, length 62 mm, Matoaka Cottages, St. Leonard, Calvert County; K,L = *Trisecphora smithae* (Petuch, 1988), length 78 mm, Sandgates, St. Mary's County, Patuxent River; M = *Ecphora harasewychi* Petuch, 1989, length 32 mm, Sandgates, St. Mary's County, Patuxent River; N = *Volutifusus conradianus* (Martin, 1904), length 57 mm, Matoaka Cottages, St. Leonard, Calvert County.

A
B
C
D
E
F
G
H
I
J
K
L
M
N
O

subspecies, the oyster *Crassostrea* cf. *carolinensis* (Conrad, 1862), the lucinid *Lucinoma contracta* (Say, 1824), and the small venerid *Callocardia sayana* (Conrad, 1833). Several small, poorly preserved bivalves are also frequently abundant in the Shattuck Zone 17 clays, with the most common species being the leptonid *Aligena aequata* (Conrad, 1843).

SHELLS OF THE DRUMCLIFF MEMBER

The Drumcliff Member (Shattuck Zone 18), like Shattuck Zone 10 in the Calvert Formation, contains the richest and most diverse molluscan fauna found in the Choptank Formation. As in all the Choptank members, the ecphoras are abundant and are the most conspicuous component of the gastropod fauna. Five ecphoras occur together in Shattuck Zone 18, and these were the dominant predators of the rich bivalve fauna. Of these, *Ecphora meganae* Ward and Gilinsky, 1988 (Figure 5.2E,F) is the most common species. The Shattuck Zone 18 descendant of *Ecphora sandgatesensis, E. meganae,* differs from its St. Leonard ancestor in being a larger, more inflated, and less cylindrical shell with thicker ribs that are broader on the edges and more "T"-shaped in cross section. The distinctive upward-turning shoulder rib of *E. sandgatesensis* is also present on *E. meganae,* but only on the early whorls of the adult shell.

The other common ecphora in Shattuck Zone 18 is the thin-ribbed *Ecphora (Planecphora) choptankensis* Petuch, 1988 (Figure 5.2C,D), the descendant of *E. (Planecphora) vokesi* from Shattuck Zone 17. This delicate, striking species with bladelike ribs differs from the older species in being a more cylindrical, less inflated shell with thinner ribs that project farther from the shell body. The umbilicus of *E. (Planecphora) choptankensis* is also much narrower than the characteristic wide, flaring umbilicus of *E. (Planecphora) vokesi. Ecphora (Planecphora) choptankensis* is most frequently encountered at the base of Drumcliff, St. Mary's County, on the Patuxent River and also around Matoaka Cottages, St. Leonard, Calvert County. *Ecphora (Planecphora) choptankensis* is sometimes confused with the sympatric *Ecphora meganae,* but can be readily separated from that species by consistently having much thinner, bladelike ribs that are rounded on their edges and by lacking the upward-turning shoulder rib on the early whorls.

Three other ecphoras are also found along with *Ecphora meganae* and *Ecphora (Planecphora) choptankensis* but are not as common and are only infrequently collected.

FIGURE 5.2 (Facing page) Index fossils for the Drumcliff Member of the Choptank Formation (Shattuck Zone 18). A,B = *Ecphora rikeri* Petuch, 1988, length 80 mm, between Matoaka Cottages, St. Leonard and Governor Run, Calvert County; C,D = *Ecphora (Planecphora) choptankensis* Petch, 1988, length 62 mm, between Matoaka Cottages and Governor Run, Calvert County; E,F = *Ecphora meganae* Ward and Gilinsky, 1988, length 69 mm, between Matoaka Cottages and Governor Run, Calvert County; G,H = *Chesathais lindae* Petuch, 1988, length 59 mm, between Matoaka Cottages and Governor Run, Calvert County; I = *Torcula dianeae* (Ward, 1992), length 35 mm, lower bed at Drumcliff, St. Mary's County, Patuxent River; J = *Glyphis nassula* (Conrad, 1845), length 28 mm, lower bed at Drumcliff, St. Mary's County, Patuxent River; K = *Calverturris schmidti* Petuch, 1993, length (incomplete, spire only) 23 mm, lower bed at Drumcliff, St. Mary's County, Patuxent River; L = *Astarte tisphila* Glenn, 1904, length 28 mm, between Matoaka Cottages and Governor Run, Calvert County; M = *Calliostoma aphelium* Dall, 1892, width 18 mm, between Matoaka Cottages and Governor Run, Calvert County; N = *Trisecphora patuxentia* (Petuch, 1988), length 35 mm, Sandgates, St. Mary's County, Patuxent River; O = *Trisecphora patuxentia* (Petuch, 1988), length 75 mm, Sandgates, St. Mary's County, Patuxent River.

A
B
C
D
E
F
G
H
I
J
K
L
M
N

Principal among these is the largest ecphora of Shattuck Zone 18, the rhomboidal-shaped *Ecphora rikeri* Petuch, 1988 (Figure 5.2A,B). This distinctive species is the descendant of the Shattuck Zone 17 *Ecphora harasewychi* and differs from that species in being a much larger, more elongated shell with a proportionally higher spire. The ribs of *E. rikeri* are also thinner and less developed than those of its ancestor, resembling thick cords adhering directly to the body whorl. The entire shell surface of *E. rikeri* is also covered with fine, raised spiral threads. Although Ward (1992) synonymized *E. rikeri* with his *E. meganae,* the two sympatric species differ greatly from each other. *Ecphora rikeri* is a much larger, more elongated shell with a proportionally higher spire and much more sloping shoulder whorls. The ribs of the two species also differ consistently, with *E. meganae* having thicker ribs that are flattened on the edges, while *E. rikeri* has thinner, much more rounded ribs that resemble cords. The ribs of *E. rikeri* are also proportionally farther apart than are the more crowded ribs of *E. meganae*.

Occurring along with *Ecphora rikeri* is *Chesathais lindae* Petuch, 1988 (Figure 5.2G,H), one of the rarest and most sought-after species from Shattuck Zone 18. This descendant of the St. Leonard *C. donaldasheri* differs from its ancestor in being a much larger, more elongated shell with a higher spire. The only three-ribbed ecphora in Shattuck Zone 18, *Trisecphora patuxentia* (Petuch, 1988) (Figure 5.2N,O), is the direct ancestor of *T. smithae* from Shattuck Zone 17. This unusual species, which is most commonly collected near Sandgates and Queen's Cliff along the Patuxent River, differs from the ancestral *T. smithae* in being a smaller, more cylindrical and less-inflated shell with thinner ribs that project farther from the body whorl. *Trisecphora patuxentia* also has undulating, wavy ribs, a characteristic feature that sets it aside from the other three-ribbed species.

The Drumcliff Member also contains a large number of distinctive gastropods that can be used as index fossils. Some of these include the keyhole limpet *Glyphis nassula* (Conrad, 1845) (Figure 5.2J), the trochid *Calliostoma aphelium* Dall, 1892 (Figure 5.2M), the small turritellid *Torcula dianeae* (Ward, 1992) (Figure 5.2I), the small naticid moon snail *Euspirella hemicrypta* (Gabb, 1860) (Figure 5.3H), the busycon whelks *Sycopsis lindae* Petuch, 1988 (Figure 5.3B) and *Busycotypus choptankensis* Petuch, 1993 (Figure 5.3N),

FIGURE 5.3 (Facing page) Index fossils for the Drumcliff Member of the Choptank Formation (Shattuck Zone 18). A = *Lucinoma contracta* (Say, 1824), length 40 mm, between Matoaka Cottages and Governor Run, Calvert County; B = *Sycopsis lindae* Petuch, 1988, length 67 mm, lower bed at Drumcliff, St. Mary's County, Patuxent River; C = *Mariaturricula rugata* (Conrad, 1862), length 42 mm, lower bed at Drumcliff, St. Mary's County, Patuxent River; D = *Chesapecten nephrens* Ward and Blackwelder, 1975, length 98 mm, between Matoaka Cottages and Governor Run, Calvert County; E = *Isognomon (Hippochaeta) torta* (Say, 1820), length 122 mm, between Matoaka Cottages and Governor Run, Calvert County; F = *Ventrilia biplicifera* (Conrad, 1841), length 42 mm; G = *Pleiorhytis calvertensis* (Dall, 1900), length 52 mm, between Matoaka Cottages and Governor Run, Calvert County; H = *Euspirella hemicrypta* (Gabb, 1860), length 11 mm, between Matoaka Cottages and Governor Run, Calvert County; I = *Marianarona gernanti* Petuch and Drolshagen, new species (see Systematic Appendix), holotype, length 12 mm, lower bed at Drumcliff, St. Mary's County, Patuxent River; J = *Macrocallista marylandica* (Conrad, 1833), length 84 mm, between Matoaka Cottages and Governor Run, Calvert County; K = *Bicorbula drumcliffensis* (Oleksyshyn, 1960), length 26 mm, lower bed at Drumcliff, St. Mary's County, Patuxent River; L = *Glossus marylandica* (Schoonover, 1941), length 54 mm, between Matoaka Cottages and Governor Run, Calvert County; M = *Marvacrassatella turgidula* (Conrad, 1843), length 68 mm, between Matoaka Cottages and Governor Run, Calvert County; N = *Busycotypus choptankensis* Petuch, 1993, holotype, length 192 mm, lower bed at Drumcliff, St. Mary's County, Patuxent River.

A
B
C
D
E
F
G
H
I
J
K
L
M
N

the nutmeg shells *Cancellariella neritoidea* (Martin, 1904), *Marianarona gernanti* Petuch and Drolshagen, new species (see Systematic Appendix), and *Ventrilia biplicifera* (Conrad, 1841) (Figure 5.3F), and the turrids *Calverturris schmidti* Petuch, 1993 and *Mariaturricula rugata* (Conrad, 18620 (Figure 5.3C).

The bivalve fauna of Shattuck Zone 18 is equally as impressive as the gastropod fauna and contains many large, prominent species. Most frequently encountered is the large scallop *Chesapecten nephrens* Ward and Blackwelder, 1975 (Figure 5.3D), which often occurs in extensive beds, and the giant pearly oyster *Isognomon (Hippochaeta) torta* (Say, 1820) (Figure 5.3E), which often occurs in large reeflike biohermal clumps much like those seen in Shattuck Zone 17. Other characteristic Drumcliff bivalves include the ark shell *Dallarca elnia* (Glenn, 1904), the crassatellid *Marvacrassatella turgidula* (Conrad, 1843) (Figure 5.3M), the astartid *Astarte tisphila* Glenn, 1904 (Figure 5.2L), the psammobiid *Pleiorhytis calvertensis* (Dall, 1900) (Figure 5.3G), the lucinid *Lucinoma contracta* (Say, 1824) (Figure 5.3A), the corbulid *Bicorbula drumcliffensis* (Oleksyshyn, 1960) (Figure 5.3K), the glossid *Glossus marylandica* (Schoonover, 1941) (Figure 5.3L), the cockle shell *Chesacardium blackwelderi* Ward, 1992, and the venerid *Macrocallista marylandica* (Conrad, 1833).

SHELLS OF THE BOSTON CLIFFS MEMBER

Like Shattuck Zone 17, the molluscan fauna of the Boston Cliffs Member (Shattuck Zone 19) is impoverished and contains only one-third the number of species found in Shattuck Zone 18. The Boston Cliffs Member marks the end of Choptank deposition, and Shattuck Zone 19 contains the last species of many Calvert–Choptank lineages. Such groups as *Patuxentrophon, Chesathais,* and *Trisecphora* make their last appearance in Shattuck Zone 19 and their disappearance is a good biostratigraphic marker for the Choptank–St. Mary's formational boundary.

As in the Drumcliff Member, five species of ecphoras are present in the Boston Cliffs member, but none are as abundant as the species found in Shattuck Zones 17 and 18. Of the Boston Cliffs ecphoras, the most frequently encountered species include the heavy, thick-ribbed *Ecphora williamsi* Ward and Gilinsky, 1988 (Figure 5.4A,B) and the fragile, thin-ribbed *Ecphora (Planecphora) delicata* Petuch, 1989 (Figure 5.4C,D). Even though Ward and Gilinsky (1988) considered *E. williamsi* to be a subspecies and descendant of the Shattuck Zone 18 *E. meganae,* this distinctive species, with its thick, rounded, and cordlike ribs, actually appears to be a descendant of the Shattuck Zone 16 *E. chesapeakensis* (see Chapter 3). The last *Planecphora* species in Maryland, *E. delicata,* is the descendant of the Shattuck Zone 18 *E. (Planecphora) choptankensis* and differs from that species in being

FIGURE 5.4 (Facing page) Index fossils for the Boston Cliffs Member of the Choptank Formation (Shattuck Zone 19). A,B = *Ecphora williamsi* Ward and Gilinsky, 1988, length 57 mm; C = *Ecphora (Planecphora) delicata* Petuch, 1989, length 39 mm; E = *Chesapecten monicae* Ward, 1992, length 36 mm; F = *Chesathais drumcliffensis* Petuch, 1989, length 32 mm; G = *Patuxentrophon patuxentia* (Martin, 1904), length 18 mm; H = *Mariacolpus octonaria* (Conrad, 1863), length 27 mm; I = *Conradconfusus patuxentensis* (Petuch, 1993), length 74 mm; J,K = *Ecphora amyae* Petuch and Drolshagen, new species (see Systematic Appendix), holotype, length 37 mm; L = *Trisecphora shattucki* (Petuch, 1989), length 25 mm; M = *Astarte obruta* Conrad, 1834, length 27 mm; N = *Marvacrassatella marylandica* (Conrad, 1832), length 70 mm. All specimens were collected from the upper bed at Drumcliff, St. Mary's County, on the Patuxent River.

a smaller, more vase-shaped shell with much thinner ribs that project much farther from the body whorl. *Ecphora (Planecphora) delicata* also has a characteristic upward-pointing shoulder rib, producing distinctly cupped, deeply canaliculate spire whorls and a much smaller and narrower umbilicus.

The rarest ecphora in Shattuck Zone 19 is *Chesathais drumcliffensis* Petuch, 1989 (Figure 5.4F), the descendant of the Shattuck Zone 18 *C. lindae.* Although similar to *C. lindae* in general appearance, *C. drumcliffensis* differs in being a smaller, more inflated shell with a lower spire. This is the last species of *Chesathais.* Equally rare is the real descendant of *Ecphora meganae, E. amyae* Petuch and Drolshagen, new species (Figure 5.4J,K) (Ward and Gilinsky incorrectly state that the rounded-ribbed *E. williamsi* is the descendant of the flat-ribbed *E. meganae;* see Systematic Appendix). This unusual species retains the upward-pointing shoulder rib of *E. meganae,* but differs in having thicker, lower ribs and in being more cylindrical in shape. *Ecphora amyae* differs from the sympatric *E. williamsi* in having ribs that are "T"-shaped in cross section with a single deep groove, not rounded with 2 to 4 fine incised threads. Another uncommon Boston Cliffs ecphora is *Trisecphora shattucki* (Petuch, 1989) (Figure 5.4L), the last of the three-ribbed ecphoras. This small species, although descending from *T. smithae* and *T. patuxentia,* has thicker ribs than its ancestors.

Other characteristic Shattuck Zone 19 gastropod index fossils include the turritellids *Mariacolpus octonaria* (Conrad, 1863) (Figure 5.4H) and *Torcula terebriformis* (Dall, 1892), the trophonine muricid *Patuxentrophon patuxentia* (Martin, 1904) (Figure 5.4G), the fasciolariid *Conradconfusus patuxentensis* (Petuch, 1993) (Figure 5.4I), and the large busycon whelk *Busycotypus choptankensis* Petuch, 1993. The Shattuck Zone 19 bivalve fauna is equally small but contains several important index fossils, including the small and rare scallop *Chesapecten monicae* Ward, 1992 (Figure 5.4E), the crassatellid *Marvacrassatella marylandica* (Conrad, 1832) (Figure 5.4N), the cockle shell *Chesacardium patuxentia* (Glenn, 1904) (*C. vostreysi* Ward, 1992 is a synonym), and the astartid *Astarte obruta* Conrad, 1834 (Figure 5.4M).

SYSTEMATIC LIST OF CHOPTANK MOLLUSKS

The following is a list of the more important index fossils of the Choptank Formation, arranged systematically and by Shattuck Zone. For illustrations of micromollusks and tiny species, see Martin (1904) and Glenn (1904).

CHOPTANK FORMATION

St. Leonard Member

Shattuck Zone 17

Gastropoda

Trochidae

Calliostoma species (*C. philanthropus* complex)

Turritellidae

Mariacolpus octonaria subspecies

Naticidae

Euspira species (*E. heros* complex)

Muricidae

Chesathais donaldasheri Petuch, 1989 (Figure 5.1C)

Ecphora harasewychi Petuch, 1989 (Figure 5.1M)

Ecphora sandgatesensis Petuch, 1989 (Figure 5.1A,B)

Ecphora (Planecphora) vokesi Petuch, 1988 (Figure 5.1I,J)

Ecphorosycon lindajoyceae (Petuch, 1993) (Figure 5.1E,F)

Trisecphora smithae (Petuch, 1988) (Figure 5.1K,L)

Stephanosalpinx candelabra Petuch, 1988 (Figure 5.1G,H)

Nassariidae

Ilyanassa gubernatoria (Martin, 1904)

Volutidae

Volutifusus conradianus (Martin, 1904) (Figure 5.1N)

Bivalvia

Arcidae

Dallarca cf. *staminea* (Say, 1832)

Pectinidae

Chesapecten nephrens subspecies

Ostreidae

Crassostrea cf. *carolinensis* (Conrad, 1862)

Isognomonidae

Isognomon (Hippochaeta) torta (Say, 1820)

Lucinidae

Lucinoma contracta (Say, 1824)

Leptonidae

Aligena aequata (Conrad, 1843)

Veneridae

Callocardia sayana (Conrad, 1833)

Mercenaria cuneata (Conrad, 1867) (Figure 5.1D)

(**Author's note:** In Petuch (2004:101–104), these species were incorrectly designated as coming from Shattuck Zone 16. This should now be emended to state that they came from Shattuck Zone 17.)

Drumcliff Member

Shattuck Zone 18

Gastropoda

Fissurellidae

Glyphis nassula (Conrad, 1845) (Figure 5.2J)

Trochidae

Calliostoma aphelium Dall, 1892 (Figure 5.2M)

Calliostoma species (*C. philanthropus* complex)

Turritellidae

Mariacolpus octonaria subspecies

Torcula dianeae Ward, 1992 (Figure 5.2I)

Crepidulidae

Crucibulum multilineatum (Conrad, 1842)

Naticidae
Euspira species (*E. heros* complex)
Euspirella hemicrypta (Gabb, 1860) (Figure 5.3H)
Muricidae
Chesathais lindae Petuch, 1988 (Figure 5.2G,H)
Ecphora meganae Ward and Gilinsky, 1988 (Figure 5.2E,F)
Ecphora rikeri Petuch, 1988 (Figure 5.2A,B)
Ecphora (Planecphora) choptankensis Petuch, 1988 (Figure 5.2C,D)
Trisecphora patuxentia (Petuch, 1988) (Figure 5.2N,O)
Busyconidae
Busycotypus choptankensis Petuch, 1993 (Figure 5.3N)
Sycopsis lindae Petuch, 1988 (Figure 5.3B)
Nassariidae
Ilyanassa peraltoides (Martin, 1904)
Ilyanassa trivittatoides (Whitfield, 1894)
Cancellariidae
Cancellariella neritoidea (Martin, 1904)
Marianarona gernanti Petuch and Drolshagen, new species
Mariasveltia patuxentia (Martin, 1904)
Ventrilia biplicifera (Conrad, 1841) (Figure 5.3F)
Turridae
Calverturris schmidti Petuch, 1993 (Figure 5.2K)
Hemipleurotoma species
Mariaturricula rugata (Conrad, 1862) (Figure 5.3C)
Sediliopsis patuxentia (Martin, 1904)
Transmariaturris patuxentia (Martin, 1904)
Bivalvia
Arcidae
Dallarca elnia (Glenn, 1904)
Pectinidae
Chesapecten nephrens Ward and Blackwelder, 1975 (Figure 5.3D)
Christinapecten marylandica (Wagner, 1839)
Stralopecten skiptonensis (Mansfield, 1939)
Pinnidae
Atrina harrisi subspecies
Anomiidae
Anomia species
Isognomonidae
Isognomon (Hippochaeta) torta (Say, 1820) (Figure 5.3E)
Crassatellidae
Marvacrassatella turgidula (Conrad, 1843) (Figure 5.3M)
Astartidae
Astarte tisphila Glenn, 1904 (Figure 5.2L)
Carditidae
Carditamera producta Conrad, 1843

Lucinidae
Lucinoma contracta (Say, 1824) (Figure 5.3A) (also found in the Calvert Formation)
Psammobiidae
Pleiorhytis calvertensis (Dall, 1900) (Figure 5.3G)
Diplodontidae
Diplodonta shilohensis Dall, 1900
Leptonidae
Aligena aequata (Conrad, 1843)
Corbulidae
Bicorbula drumcliffensis (Oleksyshyn, 1960) (Figure 5.3K)
Caryocorbula cuneata (Harris, 1896)
Caryocorbula subcontracta (Whitfield, 1894)
Glossidae
Glossus marylandica (Schoonover, 1941) (Figure 5.3L)
Hiatellidae
Panopea goldfussi Wagner, 1839
Myidae
Mya producta Conrad, 1838
Cardiidae
Chesacardium blackwelderi Ward, 1992
Laevicardium species
Veneridae
Dosinia blackwelderi Ward, 1992
Macrocallista marylandica (Conrad, 1833) (Figure 5.3J)
Mercenaria cuneata subspecies
Mactridae
Florametis biplicata (Conrad, 1834)
Tellindae
Macoma lens (Conrad, 1843)
Tellina declivis Conrad, 1834

(**Author's note:** In Petuch (2004:105–107) these species were incorrectly designated as coming from Shattuck Zone 17. This should now be emended to state that they came from Shattuck Zone 18.)

Boston Cliffs Member
Shattuck Zone 19
Gastropoda
Trochidae
Calliostoma species (*C. philanthropus* complex)
Turritellidae
Mariacolpus octonaria (Conrad, 1863) (Figure 5.4H)
Torcula terebriformis (Dall, 1892)
Vermetidae
Serpulorbis granifera subspecies

Naticidae
  Euspira species (*E. heros* complex)
Muricidae
  Chesathais drumcliffensis Petuch, 1989 (Figure 5.4F)
  Ecphora amyae Petuch and Drolshagen, new species (Figure 5.4J,K)
  Ecphora williamsi Ward and Gilinsky, 1988 (Figure 5.4A,B)
  Ecphora (Planecphora) delicata Petuch, 1989 (Figure 5.4C,D)
  Trisecphora shattucki (Petuch, 1989) (Figure 5.4L)
  Patuxentrophon patuxentia (Martin, 1904) (Figure 5.4G)
Fasciolariidae
  Conradconfusus patuxentensis (Petuch, 1993) (Figure 5.4I)
Busyconidae
  Busycotypus choptankensis Petuch, 1993
Volutidae
  Volutifusus conradianus subspecies
Terebridae
  Laevihastula patuxentia (Martin, 1904)
Bivalvia
Arcidae
  Dallarca staminea (Say, 1832)
Noetiidae
  Striarca cf. *centenaria* (Say, 1824)
Pectinidae
  Chesapecten monicae Ward, 1992 (Figure 5.4E)
Isognomonidae
  Isognomon (Hippochaeta) torta (Say, 1820)
Crassatellidae
  Marvacrassatella marylandica (Conrad, 1832) (Figure 5.4N)
Cardiidae
  Chesacardium patuxentia (Glenn, 1904)
Astartidae
  Astarte obruta Conrad, 1834 (Figure 5.4M)

6 Fossils of the St. Mary's Formation

Of the five Shattuck Zones found within the St. Mary's Formation, four contain well-preserved molluscan fossils. These include Shattuck Zone 20 (the Conoy Member), Shattuck Zones 21 and 22 (the Little Cove Point Member), and Shattuck Zone 24 (the Windmill Point Member). The prominent sandstone ledge along the St. Mary's River (the Chancellor Point Sandstone), separated from the underlying Shattuck Zone 24 by a major unconformity, also contains some well-preserved mollusks (see Chapter 2). Of all these units, the Windmill Point Member contains the richest and most diverse molluscan fauna, with the largest number of species found in the entire Chesapeake Miocene.

SHELLS OF THE CONOY MEMBER

The oldest member, the Conoy Member (Shattuck Zone 20), contains only a small number of mollusks, with most in a bad state of preservation. These usually include partially dissolved specimens of the turritellid *Mariacolpus covepointensis* subspecies, the naticid moon snail *Euspira* species, and fragments of the scallop *Chesapecten* species. The only well-preserved mollusk in the Conoy Member is the heavily constructed muricid *Ecphora conoyensis* Petuch, 2004 (Figure 6.1G,H). This small, thick, and heavy species is the descendant of Shattuck Zone 19 *Ecphora williamsi* Ward and Gilinsky, 1988 and differs from its Choptank ancestor in having ribs that are less rounded in cross section and are more flattened on their edges. The two species share the same low rounded rib type and globose, compact shell shape. Besides being the only ecphora found in the Conoy Member, *Ecphora conoyensis* also represents the perfect morphological intermediate between the *Ecphora calvertensis-williamsi* species group and the larger, more advanced forms of the *Ecphora gardnerae* species group.

SHELLS OF THE LITTLE COVE POINT MEMBER

The Little Cove Point Member (Shattuck Zones 21, 21, and 23) was deposited during the warming marine climate of the late Serravallian Age. This more moderate temperature led to the establishment of a much richer molluscan fauna than that which existed during Conoy Member time. The lower Little Cove Point bed, Shattuck Zone 21, contains an impoverished but still interesting fauna with several statigraphically restricted species. The shell beds from this unit are scattered and only locally developed, with the best exposures and collecting areas being located about two miles north of Little Cove Point, Calvert County. Some of the more important of the Shattuck Zone 21 gastropod index fossils include the muricids *Scalaspira vokesae* Petuch, 1988 (Figure 6.1E) and *Ecphora asheri* Petuch, 1988 (with five ribs; Figure 6.1A,B), the pisaniiform buccinid *Celatoconus asheri* Petuch, 1988 (Figure 6.1F), and the nutmeg shell *Marianarona lindae* Petuch and Drolshagen, new

A
B
C
D
E
F
G
H
I
J
K
L
M

species, (Figure 6.1I) (see Systematic Appendix). Two new important index bivalves are also commonly collected in Shattuck Zone 21, including the ark shell *Dallarca covepointensis* Petuch and Drolshagen, new species, (Figure 6.1M) and the venerid *Mercenaria lindae* Petuch and Drolshagen, new species, (Figure 6.1K) (see Systematic Appendix). Several long-lived, widespread St. Mary's mollusks are also present in Shattuck Zone 21, including gastropods such as the naticid *Sinum fragilis* (Conrad, 1830) (Figure 6.1L), the terebrid *Laevihastula simplex* (Conrad, 1830) (Figure 6.1D), the turrid *Hemipleurotoma communis* (Conrad, 1830) (Figure 6.1C), and the venerid bivalve *Dosinia thori* Ward, 1992 (Figure 6.1J).

The largest gastropod of Shattuck Zone 21, the five-ribbed *Ecphora asheri,* is the descendant of the Conoy Member *Ecphora conoyensis.* This distinctive species retains the rounded, globose shell shape and low, cordlike ribs of *E. conoyensis* but differs in being a much larger shell with five equal-sized ribs, not four. Some specimens of *Ecphora asheri* are highly ornate, with numerous large spiral cords in between the five main ribs. This large Shattuck Zone 21 muricid is the most highly ornamented ecphora known from the Chesapeake Miocene.

The warming trend at the Serravallian–Tortonian boundary is reflected by the explosion of large mollusks that appear suddenly in Shattuck Zone 22. Typical of some of these abundant large species are the gastropods *Ecphora germonae* Ward and Gilinsky, 1988 (Figure 6.2H,I), the busycon whelks *Coronafulgur chesapeakensis* (Petuch, 1988) (Figure 6.2A), *Busycotypus martini* Petuch and Drolshagen, new species (Figure 6.2L) (see Systematic Appendix), *Sycopsis carinatum* (Conrad, 1862) (Figure 6.2J), and *Turrifulgur covepointensis* Petuch, 2004 (Figure 6.2D), and the fasciolariid *Conradconfusus chesapeakensis* (Petuch, 1988) (Figure 6.2B), and bivalves such as the scallop *Chesapecten covepointensis* Ward, 1992 (Figure 6.2F), the cockle shell *Chesacardium eschelmani* Ward, 1992 (Figure 6.2M), the mactrid *Mactrodesma subponderosa* (d'Orbigny, 1852) (Figure 6.2C), and the venerids *Mercenaria tetrica* (Conrad, 1838) (Figure 6.2N), *Dosinia thori* Ward, 1992, and *Clementia inoceriformis* (Wagner, 1839). Numerous smaller mollusks are also abundant in Shattuck Zone 22 sediments, including gastropods such as the turritellids *Mariacolpus covepointensis* Petuch, 2004, which forms thickly packed beds (Figure 6.2G),

FIGURE 6.1 **(Facing page)** Index fossils for the Conoy Member (Shattuck Zone 20) and lower Little Cove Point Member (Shattuck Zone 21) of the St. Mary's Formation. A,B = *Ecphora asheri* Petuch, 1988, length 75 mm, Shattuck Zone 21, north side of Little Cove Point, Calvert County; C = *Hemipleurotoma communis* (Conrad, 1830), length 23 mm, Shattuck Zone 21, north side of Little Cove Point, Clavert County; D = *Laevihastula simplex* (Conrad, 1830), length 36 mm, Shattuck Zone 21, north side of Little Cove Point, Calvert County; E = *Scalaspira vokesae* Petuch, 1988, length 18 mm, Shattuck Zone 21, north side of Little Cove Point, Calvert County; F = *Celatoconus asheri* Petuch, 1988, length 20 mm, Shattuck Zone 21, north side of Little Cove Point, Calvert County; G,H = *Ecphora conoyensis* Petuch, 2004, length 51 mm, Shattuck Zone 20, 3 miles north of Little Cove Point, Calvert County; I = *Marianarona lindae* Petuch and Drolshagen, new species (see Systematic Appendix), holotype, length 17 mm, Shattuck Zone 21, north side of Little Cove Point, Calvert County; J = *Dosinia thori* Ward, 1992, length 64 mm, Shattuck Zone 21, 2 miles north of Little Cove Point, Calvert County; K = *Mercenaria lindae* Petuch and Drolshagen, new species (see Systematic Appendix), holotype, length 65 mm, ½ mile north of Little Cove Point, Calvert County; L = *Sinum fragilis* (Conrad, 1830), length 32 mm, north side of Little Cove Point, Calvert County; M = *Dallarca covepointensis* Petuch and Drolshagen, new species (see Systematic Appendix), holotype, length 48 mm, ½ mile north of Little Cove Point, Calvert County.

A
B
C
D
E
F
G
H
I
J
K
L
M
N

and *Torcula bohaskai* Ward, 1992, the muricid *Urosalpinx subrusticus bartoni* Petuch and Drolshagen, new species (Figure 6.2E) (see Systematic Appendix), and the buccinid *Bulliopsis marylandica* (Conrad, 1862) (Figure 6.2K), and bivalves such as the lucinid *Parvilucina crenulata* (Conrad, 1840) and the mactrid *Mactra clathrodon* Lea, 1833.

The single Shattuck Zone 22 ecphora, *Ecphora germonae,* is the descendant of the Shattuck Zone 21 *E. asheri* and retains its large size. *Ecphora germonae* differs from its ancestor in being a less inflated and more cylindrical shell with only four ribs, not five, and in having a longer and narrower siphonal canal. The ribs of *Ecphora germonae* are also more flattened along their edges, are distinctly "T"-shaped in cross section, and project out much farther from the body whorl. The differences in the rib structures of *E. germonae* and *E. asheri* are readily noticeable when comparing the edges of their apertures (shown on Figure 6.1B and Figure 6.2I).

SHELLS OF THE WINDMILL POINT MEMBER

With the establishment of a warm marine climate in the early Tortonian Age, the Salisbury Sea again housed an extremely rich molluscan fauna, not seen since the Langhian Age (Shattuck Zone 10 of the Calvert Formation). In the Windmill Point Member, numerous tropical groups can be seen to have migrated northward from Florida and the Carolinas into Maryland. Principal among these were tropical gastropods such as the wentletrap *Stenorhytis expansa* (Conrad, 1842) (Figure 6.3G), the tonnid *Eudolium marylandicum* Petuch, 1988 (Figure 6.4A,B; first appearance of the family Tonnidae in Maryland), the cassid *Phalium (Mariacassis) caelatum* (Conrad, 1830) (Figure 6.6N; first appearance of the family Cassidae in Maryland), the muricid *Laevityphis acuticosta* (Conrad, 1830) (Figure 6.5C), the fasciolariid *Pseudaptyxis sanctaemariae* Petuch, 1988 (Figure 6.7H,I), the buccinid *Hesperisternia petuchi* Vermeij, 2006 (Figure 6.4G; first appearance of the genus *Hesperisternia* in Maryland), the mitrid *Mitra (Nebularia) mariana* Martin, 1904 (first appearance of the family Mitridae in Maryland), the marginellid *Dentimargo* species (*denticulata* complex), the cone shells *Gradiconus deluvianus* (Green, 1840) (Figure 6.3E) and *Gradiconus sanctaemariae* (Petuch, 1988) (Figure 6.3I; the first appearance of the family Conidae in Maryland), the turrids *Cymatosyrinx limatula* (Conrad, 1830) (Figure 6.7C) and *Cymatosyrinx mariana* Petuch, 1988 (Figure 6.6B), and the acteonid bubble shell *Acteon ovoides* Conrad, 1830 (Figure 6.5J). These are the farthest-known northward extensions of the ranges of the tropical–subtropical genera *Eudolium, Phalium, Hesperisternia, Dentimargo, Mitra (Nebularia),* and *Gradiconus* in the western Atlantic Ocean.

FIGURE 6.2 (Facing page) Index fossils for the upper Little Cove Point Member (Shattuck Zone 22) of the St. Mary's Formation. A = *Coronafulgur chesapeakensis* (Petuch, 1988), length 90 mm; B = *Conradconfusus chesapeakensis* (Petuch, 1988), length 80 mm; C = *Mactrodesma subponderosa* (d'Orbigny, 1852), length 76 mm; D = *Turrifulgur covepointensis* Petuch, 2004, length 68 mm; E = *Urosalpinx subrusticus bartoni* Petuch and Drolshagen, new subspecies (see Systematic Appendix), holotype, length 28 mm; F = *Chesapecten covepointensis* Ward, 1992, length 102 mm; G = *Mariacolpus covepointensis* Petuch, 2004, length 34 mm; H,I = *Ecphora germonae* Ward and Gilinsky, 1988, length 83 mm; J = *Sycopsis carinatus* (Conrad, 1862), length 69 mm; K = *Bulliopsis marylandicus* (Conrad, 1862), length 23 mm; L = *Busycotypus martini* Petuch and Drolshagen, new species (see Systematic Appendix), holotype, length 178 mm; M = *Chesacardium eschelmani* Ward, 1992, length 92 mm; N = *Mercenaria tetrica* (Conrad, 1838), length 88 mm. All specimens collected at Little Cove Point, Calvert County.

A
B
C
D
E
F
G
H
I
J
K
L
M
N

As in Shattuck Zone 22, several large carnivorous gastropods dominate the Windmill Point molluscan fauna, and include the muricid *Ecphora gardnerae* Wilson, 1987 (Figure 6.3A,B; the Maryland State Shell), the busycon whelks *Coronafulgur coronatum* (Conrad, 1840) (Figure 6.3J), *Busycotypus rugosum* (Conrad, 1843) (Figure 6.3C; the largest gastropod in the Chesapeake Miocene), *Sycopsis tuberculatum* (Conrad, 1840) (Figure 6.3F), and *Turrifulgur fusiforme* (Conrad, 1840) (Figure 6.3H), the dwarf horse conchs *Conradconfusus parilis* (Conrad, 1830) (Figure 6.3K; often with large *Chesaconcavus* barnacles attached) and *Mariafusus marylandicus* (Martin, 1904) (Figure 6.7D), and the volutes *Volutifusus mutabilis* (Conrad, 1834) (Figure 6.4L), *V. acus* Petuch, 1988 (Figure 6.4K), *V. asheri* Petuch, 1988 (Figure 6.4H), and *Scaphellopsis solitaria* (Conrad, 1830) (Figure 6.5L). The giant scallop *Chesapecten santamaria* (Tucker, 1934) (Figure 6.3N), the largest Maryland pectinid, is present in Shattuck Zone 24, but is never as abundant as its ancestor *C. covepointensis* from Shattuck Zone 22.

Of special interest in the Windmill Point Member is the presence of three large species of the volute genus *Volutifusus* Conrad, 1863, all descendants of the Choptank *V. conradianus.* The most commonly encountered of the three is *Volutifusus mutabilis,* which differs from the other two species by having a distinct shoulder and a depressed subsutural area, which gives the spire whorls a distinct stepped appearance. The shell of *V. mutabilis* is only partially covered with a thick enamel (shown on Figure 6.4L, and reflecting differences in the mantle structures of the three volute species), with the last part of the body whorl being uncovered. The rare *Volutifusus acus* is the largest of the three volutes and differs from *V. mutabilis* in being a much more elongated, slender shell with a proportionally higher spire, in having a distinctly sloping spire without a depressed subsutural area, and in lacking any columellar plications (*V. mutabilis* has two large, prominent columellar plications, while *V. acus* is completely devoid of plications, having a smooth columella; illustrated by Martin, 1904, plate 44, Figures 8 and 9). *Volutifusus acus,* unlike *V. mutabilis,* has enamel covering only the apertural area and does not have it extending onto the body whorl. The rarest of the volute species is *Volutifusus asheri,* known from only a few specimens. This small species differs from both *V. mutabilis* and *V. acus* in being proportionally broader and more inflated, in having a much lower spire, and in having a more pronounced, sharply angled shoulder. *Volutifusus asheri,* like *V. mutabilis,* has two prominent plications on the columella and is also covered by a thick layer of enamel.

Also of interest in Shattuck Zone 24 is the presence of the richest murex shell fauna found anywhere in the Chesapeake Miocene. With the exception of the large *Ecphora gardnerae,* all the other Windmill Point muricids are small species. These include the trophonines *Chesatrophon laevis* (Martin, 1904) (Figure 6.5A), *C. lindae* (Petuch, 1988) (Figure 6.5B), *C. chesapeakensis* (Martin, 1904) (Figure 6.5G), *C. harasewychi* (Petuch,

FIGURE 6.3 (Facing page) Index fossils for the Windmill Point Member (Shattuck Zone 24) of the St. Mary's Formation. A,B = *Ecphora gardnerae* Wilson, 1987, length 96 mm; C = *Busycotypus rugosum* (Conrad, 1843), length 72 mm; D = *Sediliopsis distans* (Conrad, 1862), length 17 mm; E = *Gradiconus deluvianus* (Green, 1830), length 58 mm; F = *Sycopsis tuberculatum* (Conrad, 1840), length 58 mm; G = *Stenorhytis expansa* (Conrad, 1842), length 23 mm; H = *Turrifulgur fusiforme* (Conrad, 1840), length 69 mm; I = *Gradiconus sanctaemariae* (Petuch, 1988), length 40 mm; J = *Coronafulgur coronatum* (Conrad, 1840), length 98 mm; K = *Conradconfusus parilis* (Conrad, 1832), length 106 mm; L = *Leiotrochus reclusum* (Conrad, 1830), length 24 mm; M = *Lirophora alveata* (Conrad, 1830), length 24 mm; N = *Chesapecten santamaria* (Tucker, 1934), length 135 mm. All specimens collected at Chancellor Point, St. Mary's County, on the St. Mary's River.

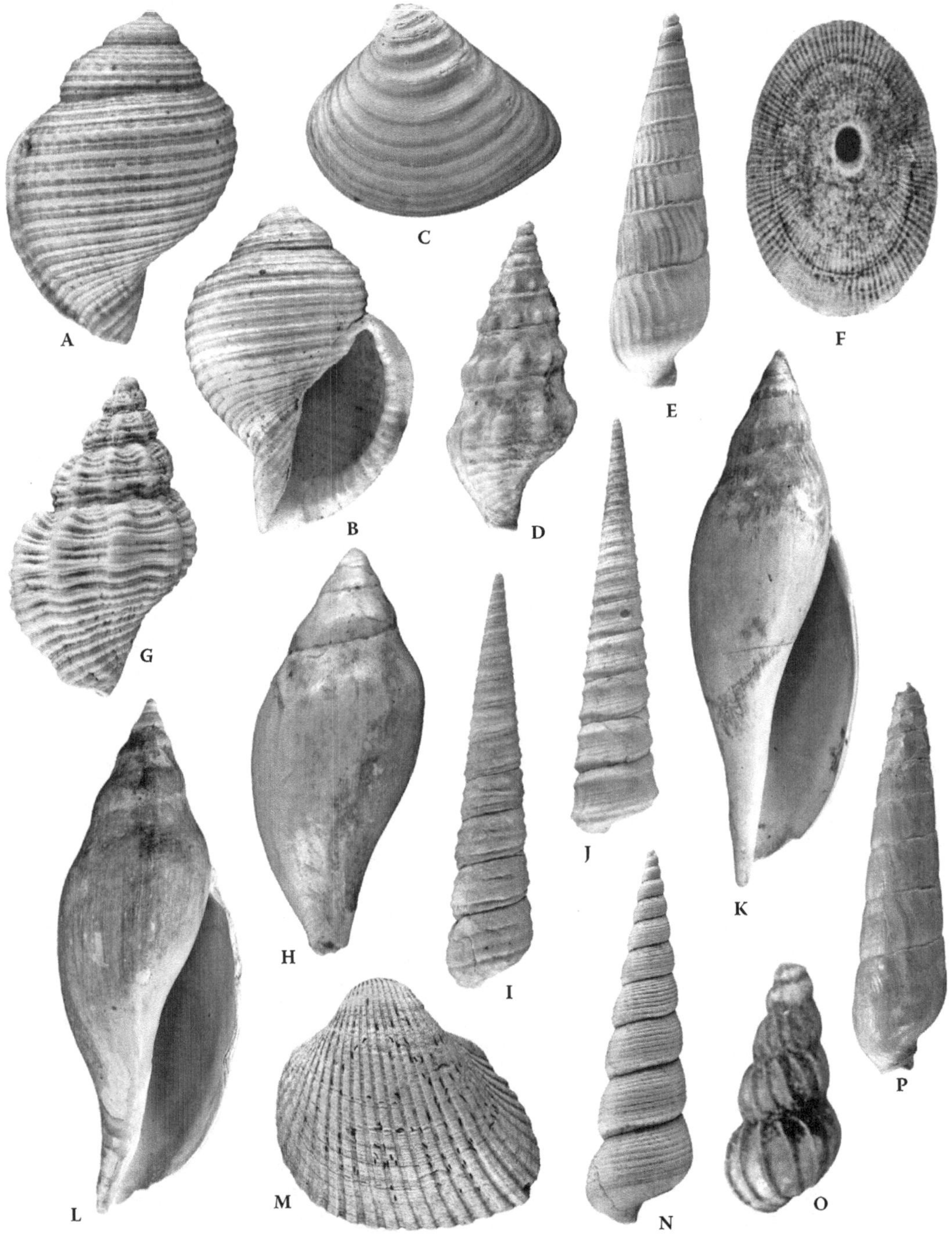
A
B
C
D
E
F
G
H
I
J
K
L
M
N
O
P

1988) (Figure 6.5M), *Lirosoma mariana* Petuch, 1988 (Figure 6.5F), and *Scalaspira harasewychi* Petuch, 1988 (Figure 6.5D), the ocenebrines *Urosalpinx subrusticus* d'Orbigny, 1852 (Figure 6.5K) and *Mariasalpinx emilyae* Petuch, 1988 (Figure 6.5I; with small labial tooth), and the typhinine *Laevityphis acuticosta* (Conrad, 1830) (Figure 6.5C).

Competing with the murex shells as drilling predators was the largest moon snail fauna of the entire Chesapeake Miocene. This radiation, comprising seven species in five genera, included the high-spired shark-eye *Neverita asheri* Petuch, 1988 (Figure 6.6F), the small, flattened *N. discula* Petuch, 1988 (Figure 6.6C), the large, inflated *Euspira interna* (Say, 1822) (Figure 6.6D), *E.* species (*E. perspectiva* complex), the small, high-spired *Euspirella santamaria* Petuch and Drolshagen, new species (Figure 6.6G) (see Systematic Appendix), the narrow umbilicus *Poliniciella marylandica* Petuch, 1988 (Figure 6.6P), and *Sinum fragilis* (Conrad, 1830). The coexistence of such a large fauna of naticid and muricid drilling predators attests to the overwhelming abundance of bivalve and gastropod prey in Shattuck Zone 24.

The biomass of polychaete worms also must have been extremely large in the Windmill Point Member, as its fauna contains the largest number of vermivorous (worm-eating) conoidean gastropods found anywhere in the Chesapeake Miocene. To date, 24 species of vermivorous gastropods coexisted during Shattuck Zone 24 time, comprising three families and 12 genera (see systematic list at the end of the chapter). Some of the more abundant and characteristic species include the previously mentioned cone shells *Gradiconus deluvianus* (Green, 1840) and *G. sanctaemariae* (Petuch, 1988) (which differs from *G. deluvianus* in being a broader, less elongated shell with a much wider shoulder and lower spire, and which lacks heavy spiral threads on the anterior end), a radiation of the smooth auger shell genus *Laevihastula* Petuch, 1988, including *Laevihastula chancellorensis* (Oleksyshyn, 1960), *L. marylandica* Petuch, 1988 (Figure 6.4P; which has small, low knobs around the suture), *L. simplex* (Conrad, 1830), and *L. sublirata* (Conrad, 1863) (Figure 6.7G), several ribbed auger shells including *Strioterebrum curvilineatum* (Conrad, 1843) (Figure 6.7E) and *S. sincerum* (Dall, 1895) (Figure 6.4E), and the turrids *Chesasyrinx rotifera* (Conrad, 1830) (Figure 6.7B), *C. mariana* (Martin, 1904) (Figure 6.7K), *Chesaclava dissimilis* (Conrad, 1830) (Figure 6.7Q), *C. quarlesi* Petuch, 1988 (Figure 6.7P), *Mariaturricula biscatenaria* (Conrad, 1830) (the largest turrid in the Chesapeake Miocene), *Nodisurculina engonata* (Conrad, 1862) (Figure 6.7N), *Mariadrillia parvoidea* (Martin, 1904) (Figure 6.6L), *Cymatosyrinx limatula* (Conrad, 1830), and *C. mariana* Petuch, 1988 (Figure 6.6B). At least six species of smaller turrids and microturrids occurred along with the larger species, including *Oenopota parva* (Conrad, 1830), *O. marylandica* (Petuch, 1988), *Sediliopsis*

FIGURE 6.4 (Facing page) Index fossils for the Windmill Point Member (Shattuck Zone 24) of the St. Mary's Formation. A,B = *Eudolium marylandicum* Petuch, 1988, length 44 mm; C = *Astarte perplana (*Conrad, 1840), length 27 mm; D = *Sediliopsis incilifera* (Conrad, 1834), 16 mm; E = *Strioterebrum sincerum* (Dall, 1895), length 28 mm; F = *Glyphis fluviamariana* (Petuch, 1988), holotype, length 26 mm; G = *Hesperisternia petuchi* Vermeij, 2006, length 23 mm; H = *Volutifusus asheri* Petuch, 1988, length 74 mm; I = *Torcula subvariabilis* (d'Orbigny, 1852), length 48 mm; J = *Torcula chancellorensis* (Oleksyshyn, 1959), length 32 mm; K = *Volutifusus acus* Petuch, 1988, length 96 mm; L = *Volutifusus mutabilis* (Conrad, 1843), length 99 mm; M = *Dallarca idonea* (Conrad, 1832), length 43 mm; N = *Mariacolpus lindajoyceae* Petuch, 2004, length 38 mm; O = *Epitonium chancellorensis* Petuch, 1988, holotype, length 4 mm; P = *Laevihastula marylandica* Petuch, 1988, length 14 mm (note low, undulating knobs around suture). All specimens were collected at Chancellor Point, St. Mary's County, on the St. Mary's River.

A
B
C
D
E
F
G
H
I
J
K
L
M
N

incilifera (Conrad, 1834) (Figure 6.4D), *S. distans* (Conrad, 1862) (Figure 6.3D), *S .angulata* (Martin, 1904), and *S. gracilis* (Conrad, 1830).

The scavenger/carnivore superfamily Buccinoidea was also well represented in Shattuck Zone 24, with two families and four genera being present. These include the buccinids *Bulliopsis quadrata* (Conrad, 1830) (Figure 6.6A), *B. integra* (Conrad, 1842) (Figure 6.6E), *B. ovata* (Conrad, 1862) (Figure 6.6H), *B. subcylindrica* (Conrad, 1866) (Figure 6.6K), *Hesperisternia petuchi* Vermeij, 2006 (Figure 6.4G; referred to as "*H. cumberlandiana*" by previous authors), *Ptychosalpinx lindae* Petuch (Figure 6.5N), and *P. pustulosus* Petuch, 1988 (Figure 6.5H), and the nassariid mud snails *Ilyanassa marylandica* (Martin, 1904) (Figure 6.6I) and *I. peralta* (Conrad, 1830) (Figure 6.6M). With four sympatric species, the genus *Bulliopsis* underwent a species radiation during Shattuck Zone 24 time, an event unique to the Windmill Point Member. After that time, only a single *Bulliopsis* species survived into the Messinian Miocene, *B. bowlerensis* Allmon, 1990 from the Eastover Formation of Virginia (see Chapter 7). Other common and characteristic Windmill Point gastropods, which can be used as Shattuck Zone 24 index fossils, include the turritellids *Mariacolpus lindajoyceae* Petuch, 2004 (Figure 6.4N; which frequently dominates the molluscan assemblage), *Torcula subvariabilis* (d'Orbigny, 1852) (Figure 6.4I), and *T. chancellorensis* (Oleksyshyn, 1959) (Figure 6.4J), the trochid *Leiotrochus reclusum* (Conrad, 1830) (Figure 6.3L), the keyhole limpet *Glyphis fluviamariana* (Petuch, 1988) (Figure 6.4F), the crepidulid cup-and-saucer shells *Crucibulum pileolum* (H.C. Lea, 1843) (Figure 6.6J) and *C. constrictum* (Conrad, 1842), the wentletraps *Epitonium chancellorensis* Petuch, 1988 (Figure 6.4O) and *E. sayanum* (Dall, 1889) (Figure 6.5E), and the nutmeg shells *Marianarona asheri* Petuch, 1988 (Figure 6.7J), *M. marylandica* Petuch, 1988 (Figure 6.7O), *M. alternata* (Conrad, 1834) (Figure 6.7F), *Mariasveltia lunata* (Conrad, 1830) (Figure 6.7M), and *Admete marylandica* (Martin, 1904) (Figure 6.7L).

Unlike the molluscan faunas of the Calvert and Choptank Formations, which are dominated by bivalves, the fauna of Shattuck Zone 24 is dominated by gastropods. These are predominantly turritellids such as *Mariacolpus lindajoyceae,* but also muricids, nassariids, and naticids. A few large bivalves are also abundant in Shattuck Zone 24, but these are often overwhelmed by the great diversity of the gastropods. Some of the more frequently encountered bivalves include the ark shell *Dallarca idonea* (Conrad, 1832) (Figure 6.4M), which formed large, densely packed beds, the scallop *Chesapecten santamaria* (Tucker, 1934), the venerids *Dosinia thori* Ward, 1992, *Lirophora alveata* (Conrad, 1831) (Figure 6.3M), and *Mercenaria tetrica* (Conrad, 1838), the astartid *Astarte perplana* Conrad, 1840, the glossid *Glossus santamaria* Ward, 1992, and the cockle shell *Chesacardium laqueatum* (Conrad, 1831).

FIGURE 6.5 **(Facing page)** Index fossils for the Windmill Point Member (Shattuck Zone 24) of the St. Mary's Formation. A = *Chesatrophon laevis* (Martin, 1904), length 16 mm; B = *Chesatrophon lindae* (Petuch, 1988), length 15 mm; C = *Laevityphis acuticosta* (Conrad, 1830), length 22 mm; D = *Scalaspira harasewychi* Petuch, length 15 mm; E = *Epitonium sayanum* (Dall, 1898), length 16 mm; F = *Lirosoma mariana* Petuch, 1988, length 25 mm; G = *Chesatrophon chesapeakensis* (Martin, 1904), length 11 mm; H = *Ptychosalpinx pustulosus* Petuch, 1988, length 27 mm; I = *Mariasalpinx emilyae* Petuch, 1988, length 28 mm; J = *Acteon ovoides* (Conrad, 1830), length 13 mm; K = *Urosalpinx subrusticus* (d'Orbigny, 1852), length 25 mm; L = *Scaphellopsis solitaria* (Conrad, 1830), length 37 mm; M = *Chesatrophon harasewychi* (Petuch, 1988), holotype, length 16 mm; N = *Ptychosalpinx lindae* Petuch, 1988, length 26 mm. All specimens were collected at Chancellor Point, St. Mary's County, on the St. Mary's River.

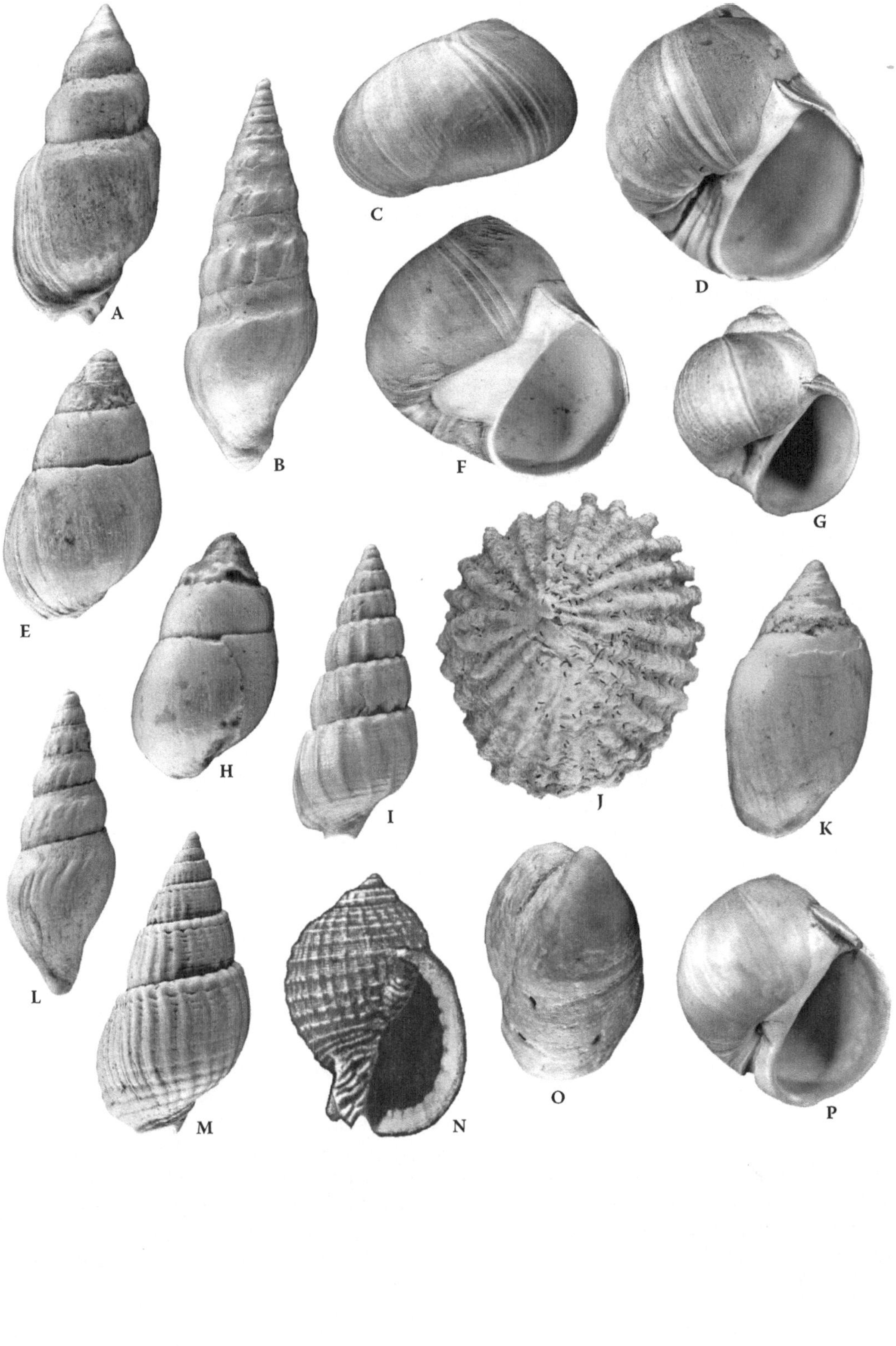
A
B
C
D
E
F
G
H
I
J
K
L
M
N
O
P

SHELLS OF THE CHANCELLOR POINT SANDSTONE

Between Chancellor Point and St. Inigoe's Creek, St. Mary's City, on the St. Mary's River, the Chancellor Point Sandstone crops out directly above Shattuck Zone 24 (see Chapter 2). On weathered surfaces, well-preserved molluscan fossils can be extracted intact, although most are in a fragmentary state. The Chancellor Point Sandstone contains a distinct and mostly undescribed fauna, and one descended from the Shattuck Zone 24 fauna. Unfortunately, the best exposures of the Chancellor Point Sandstone are now covered by riprap and piers (from adjacent housing developments), but small outcrops are occasionally exposed in erosional gullies near the tide line.

The gastropod fauna of the Chancellor Point Sandstone, although small, contains several interesting endemic species, including the busycon whelks *Busycotypus asheri* Petuch, 1988 (Figure 6.8E; with a characteristic sharp keel-like carina around the shoulder) and *Turrifulgur turriculus* Petuch, 1988 (Figure 6.8M,N; the last-living species of the genus *Turrifulgur*), the small, globose dwarf moon snail *Euspirella chancellorensis* Petuch and Drolshagen, new species (Figure 6.8C,D) (see Systematic Appendix), the small mud snail *Ilyanassa chancellorensis* Petuch and Drolshagen, new species (Figure 6.8I,J) (see Systematic Appendix), and the giant volute *Megaptygma meganucleus* Petuch, 1988 (known only from spire fragments; Petuch, 1988a). The bivalve fauna, likewise, is small, dominated by the ark shell *Dallarca chancellorensis* Petuch and Drolshagen, new species (Figure 6.8K,L) (see Systematic Appendix), the scallop *Chesapecten* species (known only from fragments), and the small venerid *Lirophora alveata* subspecies.

Three other gastropods found in the Chancellor Point Sandstone are particularly noteworthy. These include *Ecphora gardnerae angusticostata* Petuch, 1989 (Figure 6.8A,B), a small subspecies of the Windmill Point *Ecphora gardnerae,* the volute *Volutifusus marylandicus* Petuch, 1988 (Figure 6.8F), and the slender, elongated cone shell *Gradiconus asheri* (Petuch, 1988) (Figure 6.8G,H). *Ecphora gardnerae angusticostata* differs from its Shattuck Zone 24 ancestor in being a much smaller, more cylindrical shell with thinner ribs that project out farther from the body whorl. The ribs are also not as flattened on their edges nor as "T"-shaped in cross section. The impressive *Volutifusus marylandicus* is the largest species of *Volutifusus* in the Chesapeake Miocene, with some individuals reaching over 150 mm in length. This descendant of the Shattuck Zone 24 *V. mutabilis* differs from its ancestor in being a much larger, much more elongated shell with a proportionally higher spire and more sloping shoulder areas. The early whorls of *V. marylandicus* are also

FIGURE 6.6 (Facing page) Index fossils for the Windmill Point Member (Shattuck Zone 24) of the St. Mary's Formation. A = *Bulliopsis quadrata* (Conrad, 1830), length 23 mm; B = *Cymatosyrinx mariana* Petuch, 1988, length 27 mm; C = *Neverita discula* Petuch, 1988, length 18 mm; D = *Euspira interna* (Say, 1822), length 43 mm; E = *Bulliopsis integra* (Conrad, 1842), length 18 mm; F = *Neverita asheri* Petuch, 1988, length 42 mm; G = *Euspirella santamaria* Petuch and Drolshagen, new species (see Systematic Appendix), holotype, length 18 mm; H = *Bulliopsis ovata* (Conrad, 1862), length 19 mm; I = *Ilyanassa marylandica* (Martin, 1904), length 14 mm; J = *Crucibulum pileolum* (H.C. Lea, 1843), length 47 mm; K = *Bulliopsis subcylindrica* (Conrad, 1866), length 21 mm; L = *Mariadrillia parvoidea* (Martin, 1904), length 10 mm; M = *Ilyanassa peralta* (Conrad, 1830), length 22 mm; N = *Phalium (Mariacassis) caelatum* (Conrad, 1830), length 40 mm (taken from Martin, 1904, plate 40, figure 4), most often collected as broken fragments; O = *Crucibulum constrictum* (Conrad, 1842), length 21 mm, lived attached to *Chesapecten santamaria* and conformed to the shape of the scallop's ribs; P = *Poliniciella marylandica* Petuch, 1988, length 20 mm. All specimens collected at Chancellor Point, St. Mary's County, on the St. Mary's River.

A
B
C
D
E
F
G
H
I
J
K
L
M
N
O
P
Q

much more heavily ornamented than those of *V. mutabilis,* bearing large, rounded knobs on the first three whorls. One of the rarest and most desirable Chancellor Point species is the high-spired cone shell *Gradiconus asheri* (Petuch, 1988) (Figure 6.8G,H), known only from a handful of specimens. This distinctive cone shell is descended from the common Shattuck Zone 24 *G. deluvianus,* but differs in being a much more slender, elongated shell with a much higher, more projecting spire and with prominently stepped, scalariform spire whorls. *Gradiconus asheri* is the last species of cone shell to live in Maryland.

SYSTEMATIC LIST OF ST. MARY'S MOLLUSKS

The following is a list of the more important index fossils of the St. Mary's Formation, arranged systematically and by Shattuck Zone. For illustrations and descriptions of micromollusks and tiny species, see Martin (1904) and Glenn (1904). Only Shattuck Zones that contain fossils are listed.

ST. MARY'S FORMATION

Conoy Member

Shattuck Zone 20

Gastropoda

Turritellidae

Mariacolpus covepointensis subspecies

Naticidae

Euspira species (*heros* complex)

Muricidae

Ecphora conoyensis Petuch, 2004 (Figure 6.1G,H)

Bivalvia

Pectinidae

Chesapecten species

Little Cove Point Member

Shattuck Zone 21

Gastropoda

Turritellidae

FIGURE 6.7 **(Facing page)** Index fossils for the Windmill Point Member (Shattuck Zone 24) of the St. Mary's Formation. A = *Mariaturricula biscatenaria* (Conrad, 1830), length 36 mm; B = *Chesasyrinx rotifera* (Conrad, 1830), length 23 mm; C = *Cymatosyrinx limatula* (Conrad, 1830), length 24 mm; D = *Mariafusus marylandicus* (Martin, 1904), length 60 mm; E = *Strioterebrum curvilineatum* (Conrad, 1843), length 29 mm; F = *Marianarona alternata* (Conrad, 1834), length 22 mm; G = *Laevihastula sublirata* (Conrad, 1863), length 29 mm; H,I = *Pseudaptyxis sanctaemariae* Petuch, 1988, holotype, length 26 mm; J = *Marianarona asheri* Petuch, 1988, length 12 mm; K = *Chesasyrinx mariana* (Martin, 1904), length 26 mm; L = *Admete marylandica* (Martin, 1904), length 9 mm; M = *Mariasveltia lunata* (Conrad, 1830), length 14 mm; N = *Nodisurculina engonata* (Conrad, 1862), length 14 mm; O = *Marianarona marylandica* Petuch, 1988, length 12 mm; P = *Chesaclava quarlesi* Petuch, 1988, length 14 mm; Q = *Chesaclava dissimilis* (Conrad, 1830), length 16 mm. All specimens were collected at Chancellor Point, St. Mary's County, on the St. Mary's River.

A
B
C
D
E
F
G
H
I
J
K
L
M
N

Mariacolpus covepointensis subspecies
Naticidae
Euspira species (*heros* complex)
Sinum fragilis (Conrad, 1830) (Figure 6.1L)
Muricidae
Ecphora asheri Petuch, 1988 (Figure 6.1A,B)
Scalaspira vokesae Petuch, 1988 (Figure 6.1E)
Buccinidae
Celatoconus asheri Petuch, 1988 (Figure 6.1F)
Cancellariidae
Marianarona lindae Petuch and Drolshagen, new species (Figure 6.1I)
Terebridae
Laevihastula simplex (Conrad, 1830) (Figure 6.1D)
Turridae
Hemipleurotoma communis (Conrad, 1830) (Figure 6.1C)
Sediliopsis species
Bivalvia
Arcidae
Dallarca covepointensis Petuch and Drolshagen, new species (Figure 6.1M)
Pectinidae
Chesapecten cf. *covepointensis* Ward, 1992
Veneridae
Mercenaria lindae Petuch and Drolshagen, new species (Figure 6.1K)
Mercenaria tetrica subspecies
Dosinia thori Ward, 1992 (Figure 6.1J)

Shattuck Zone 22

Gastropoda
Turritellidae
Mariacolpus covepointensis Petuch, 2004 (Figure 6.2G)
Torcula bohaskai (Ward, 1992)
Naticidae
Euspira species (*heros* complex)

FIGURE 6.8 **(Facing page)** Index fossils for the Chancellor Point Sandstone. A,B = *Ecphora gardnerae angusticostata* Petuch, 1989, length 40 mm; C,D = *Euspirella chancellorensis* Petuch and Drolshagen, new species (see Systematic Appendix), holotype, length 9 mm; E = *Busycotypus asheri* Petuch, 1988, holotype, length 54 mm; F = *Volutifusus marylandicus* Petuch, 1988, holotype, length 144 mm; G,H = *Gradiconus asheri* (Petuch, 1988), length 42 mm; I,J = *Ilyanassa chancellorensis* Petuch and Drolshagen, new species (see Systematic Appendix), holotype, length 10 mm; K,L = *Dallarca chancellorensis* Petuch and Drolshagen, new species (see Systematic Appendix), holotype, length 51 mm, with sandstone matrix filling interior of valve; M = *Turrifulgur turriculus* Petuch, 1988, length 30 mm; N = *Turrifulgur turriculus* Petuch, 1988, holotype, length 32 mm. All specimens were collected in the Chancellor Point Sandstone ledges at Chancellor Point, St. Mary's City, St. Mary's County, Maryland, on the St. Mary's River.

Muricidae
Ecphora germonae Ward and Gilinsky, 1988 (Figure 6.2H,I)
Urosalpinx subrusticus bartoni Petuch and Drolshagen, new species (Figure 6.2E)
Busyconidae
Busycotypus martini Petuch and Drolshagen, new species (Figure 6.2L)
Coronafulgur chesapeakensis (Petuch, 1988) (Figure 6.2A)
Sycopsis carinatus (Conrad, 1862) (Figure 6.2J)
Turrifulgur covepointensis Petuch, 2004 (Figure 6.2D)
Fasciolariidae
Conradconfusus chesapeakensis (Petuch, 1988) (Figure 6.2B)
Buccinidae
Bulliopsis marylandica (Conrad, 1862) (Figure 6.2K)
Columbellidae
Mitrella communis (Conrad, 1830)
Volutidae
Volutifusus mutabilis subspecies
Turridae
Hemipleurotoma communis (Conrad, 1830) (Figure 6.1C)
Sediliopsis incilifera (Conrad, 1834)
Acteonidae
Acteon ovoides Conrad, 1830
Acteon cf. *shilohensis* Whitfield, 1894
Bivalvia
Arcidae
Dallarca covepointensis Petuch and Drolshagen, new species
Pectinidae
Chesapecten covepointensis Ward, 1992 (Figure 6.2F)
Mytilidae
Modiolus ducatelli Conrad, 1840
Lucinidae
Parvilucina crenulata (Conrad, 1840)
Cardiidae
Chesacardium eschelmani Ward, 1992 (Figure 6.2M)
Mactridae
Mactra clathrodon H.C. Lea, 1833
Mactrodesma subponderosa (d'Orbigny, 1852) (Figure 6.2C)
Veneridae
Clementia inoceriformis (Wagner, 1839)
Mercenaria tetrica (Conrad, 1838) (Figure 6.2N)

Windmill Point Member
Shattuck Zone 24
Gastropoda
Fissurellidae
Glyphis alticosta (Conrad, 1834)
Glyphis fluviamariana (Petuch, 1988) (Figure 6.4F)

Trochidae
- *Calliostoma distans* (Conrad, 1862)
- *Calliostoma humile* (Conrad, 1830)
- *Leiotrochus reclusum* (Conrad, 1830) (Figure 6.3L)

Turritellidae
- *Mariacolpus lindajoyceae* Petuch, 2004 (Figure 6.4N)
- *Torcula chancellorensis* (Oleksyshyn, 1959) (Figure 6.4J)
- *Torcula subvariabilis* (d'Orbigny, 1852) (Figure 6.4I)

Crepidulidae
- *Calyptraea concentricum* H.C. Lea, 1843
- *Crepidula densata* subspecies
- *Crucibulum constrictum* (Conrad, 1842) (Figure 6.6O)
- *Crucibulum pileolum* (H.C. Lea, 1843) (Figure 6.6J)

Epitoniidae
- *Epitonium chancellorensis* Petuch, 1988 (Figure 6.4O)
- *Epitonium sayanum* (Dall, 1898) (Figure 6.5E)
- *Stenorhytis expansa* (Conrad, 1842) (Figure 6.3G)

Cassidae
- *Phalium (Mariacassis) caelatum* (Conrad, 1830) (Figure 6.6N)

Tonnidae
- *Eudolium marylandicum* Petuch, 1988 (Figure 6.4A,B)

Naticidae
- *Euspira interna* (Say, 1822) (Figure 6.6D)
- *Euspira* species (*perspectiva* complex)
- *Euspirella santamaria* Petuch and Drolshagen, new species (Figure 6.6G)
- *Neverita asheri* Petuch, 1988 (Figure 6.6F)
- *Neverita discula* Petuch, 1988 (Figure 6.6C)
- *Poliniciella marylandica* Petuch, 1988 (Figure 6.6P)
- *Sinum fragilis* (Conrad, 1830)

Muricidae
- *Chesatrophon chesapeakensis* (Martin, 1904) (Figure 6.5G)
- *Chesatrophon harasewychi* (Petuch, 1988) (Figure 6.5M)
- *Chesatrophon laevis* (Martin, 1904) (Figure 6.5A)
- *Chesatrophon lindae* (Petuch, 1988) (Figure 6.5B)
- *Ecphora gardnerae* Wilson, 1987 (Figure 6.3A,B)
- *Laevityphis acuticosta* (Conrad, 1830) (Figure 6.5C)
- *Lirosoma mariana* Petuch, 1988 (Figure 6.5F)
- *Mariasalpinx emilyae* Petuch, 1988 (Figure 6.5I)
- *Scalaspira harasewychi* Petuch, 1988 (Figure 6.5D)
- *Urosalpinx subrustica* (d'Orbigny, 1852) (Figure 6.5K)

Fasciolariidae
- *Conradconfusus parilis* (Conrad, 1830) (Figure 6.3K)
- *Mariafusus marylandicus* (Martin, 1904) (Figure 6.7D)
- *Pseudaptyxis sanctaemariae* Petuch, 1988 (Figure 6.7H,I)

Busyconidae
- *Busycotypus rugosum* (Conrad, 1843) (Figure 6.3C)
- *Coronafulgur coronatum* (Conrad, 1840) (Figure 6.3J)

- *Sycopsis tuberculatum* (Conrad, 1840) (Figure 6.3F)
- *Turrifulgur fusiforme* (Conrad, 1840) (Figure 6.3H)

Buccinidae

- *Bulliopsis integra* (Conrad, 1842) (Figure 6.6E)
- *Bulliopsis ovata* (Conrad, 1862) (Figure 6.6H)
- *Bulliopsis quadrata* (Conrad, 1830) (Figure 6.6A)
- *Bulliopsis subcylindrica* (Conrad, 1866) (Figure 6.6K)
- *Celatoconus* species (fragments only)
- *Hesperisternia petuchi* Vermeij, 2006 (Figure 6.4G)
- *Ptychosalpinx lindae* Petuch, 1988 (Figure 6.5N)
- *Ptychosalpinx pustulosus* Petuch, 1988 (Figure 6.5H)

Columbellidae

- *Mitrella communis* (Conrad, 1862)

Nassariidae

- *Ilyanassa marylandica* (Martin, 1904) (Figure 6.6I)
- *Ilyanassa peralta* (Conrad, 1830) (Figure 6.6M)

Volutidae

- *Scaphellopsis solitaria* (Conrad, 1830) (Figure 6.5L)
- *Volutifusus acus* Petuch, 1988 (Figure 6.4K)
- *Volutifusus asheri* Petuch, 1988 (Figure 6.4H)
- *Volutifusus mutabilis* (Conrad, 1834) (Figure 6.4L)

Mitridae

- *Mitra (Nebularia) mariana* Martin, 1904

Marginellidae

- *Dentimargo* species (*denticulata* complex)

Cancellariidae

- *Admete marylandica* (Martin, 1904) (Figure 6.7L)
- *Marianarona alternata* (Conrad, 1834) (Figure 6.7F)
- *Marianarona asheri* Petuch, 1988 (Figure 6.7J)
- *Marianarona marylandica* Petuch, 1988 (Figure 6.7O)
- *Mariasveltia lunata* (Conrad, 1830) (Figure 6.7M)

Conidae

- *Gradiconus deluvianus* (Green, 1840) (Figure 6.3E)
- *Gradiconus sanctaemariae* (Petuch, 1988) (Figure 6.3I)

Terebridae

- *Laevihastula chancellorensis* (Oleksyshyn, 1960)
- *Laevihastula marylandica* Petuch, 1988 (Figure 6.4P)
- *Laevihastula simplex* (Conrad, 1830) (Figure 6.1D)
- *Laevihastula sublirata* (Conrad, 1863) (Figure 6.7G)
- *Strioterebrum curvilineatum* (Conrad, 1843) (Figure 6.7E)
- *Strioterebrum sincerum* (Dall, 1895) (Figure 6.4E)

Turridae

- *Chesaclava dissimilis* (Conrad, 1830) (Figure 6.7Q)
- *Chesaclava quarlesi* Petuch, 1988 (Figure 6.7P)
- *Chesasyrinx mariana* (Martin, 1904) (Figure 6.7K)
- *Chesasyrinx rotifera* (Conrad, 1830) (Figure 6.7B)
- *Cymatosyrinx limatula* (Conrad, 1830) (Figure 6.7C)

Cymatosyrinx mariana Petuch, 1988 (Figure 6.6B)
Cymatosyrinx pyramidalis (Martin, 1904)
Hemipleurotoma communis (Conrad, 1830) subspecies
Mariadrillia parvoidea (Martin, 1904) (Figure 6.6L)
Mariaturricula biscatenaria (Conrad, 1830) (Figure 6.7A)
Nodisurculina engonata (Conrad, 1862) (Figure 6.7N)
Oenopota marylandica (Petuch, 1988)
Oenopota parva (Conrad, 1830)
Sediliopsis angulata (Martin, 1904)
Sediliopsis distans (Conrad, 1862) (Figure 6.3D)
Sediliopsis gracilis (Conrad, 1830)
Sediliopsis incilifera (Conrad, 1834) (Figure 6.4D)

Acteonidae
Acteon ovoides (Conrad, 1830) (Figure 6.5J)

Bivalvia

Arcidae
Anadara arata (Say, 1824)
Dallarca idonea (Conrad, 1832) (Figure 6.4M)
Rasia andrewsi Ward, 1992

Pectinidae
Chesapecten santamaria (Tucker, 1934) (Figure 6.3N)

Veneridae
Lirophora alveata (Conrad, 1831) (Figure 6.3G)
Mercenaria tetrica (Conrad, 1838)

Mactridae
Mactra delumbis Conrad, 1832

Astartidae
Astarte perplana Conrad, 1840 (Figure 6.4C)

Glossidae
Glossus santamaria Ward, 1992

Cardiidae
Chescardium laqueatum (Conrad, 1831)

Chancellor Point Sandstone

Gastropoda

Naticidae
Euspirella chancellorensis Petuch and Drolshagen, new species (Figure 6.8C,D)

Muricidae
Ecphora gardnerae angusticostata Petuch, 1989 (Figure 6.8A,B)

Busyconidae
Busycotypus asheri Petuch, 1988 (Figure 6.8E)
Turrifulgur turriculus Petuch, 1988 (Figure 6.8M,N)

Nassariidae
Ilyanassa chancellorensis Petuch and Drolshagen, new species (Figure 6.8I,J)

Volutidae

Megaptygma meganucleus Petuch, 1988

Volutifusus marylandicus Petuch, 1988 (Figure 6.8F)

Conidae

Gradiconus asheri (Petuch, 1988) (Figure 6.8G,H)

Bivalvia

Arcidae

Dallarca chancellorensis Petuch and Drolshagen, new species (Figure 6.8K,L)

Pectinidae

Chesapecten species (known only from fragments)

Veneridae

Lirophora alveata subspecies

7 Fossils of the Eastover Formation

The Eastover Formation contains a transitional molluscan fauna, with some classic Transmarian genera that survived from St. Mary's time and with several newly evolved genera that are important taxa in the subsequent Pliocene and early Pleistocene formations of the Chesapeake Group. During the intervening cold time between the St. Mary's and Eastover Formations (mid-Tortonian), a large number of characteristic Transmarian endemic genera became extinct. Some of these included the busycon whelk genera *Sycopsis* and *Turrifulgur,* the fasciolariids *Conradconfusus* and *Pseudaptyxis,* the naticid moon snail *Poliniciella,* the cancellariids *Marianarona* and *Mariasveltia,* and the turrid genera *Mariadrillia, Chesasyrinx, Chesaclava, Nodisurculina,* and *Mariaturricula.* On the other hand, the Eastover Formation contains the first representatives of prominent Plio–Pleistocene groups such as the spiny slipper shell subgenus *Bostrycapulus,* the scallop genus *Carolinapecten,* the oyster genera *Mansfieldostrea* (see Systematic Appendix) and *Conradostrea,* and the bittersweet clam *Costaglycymeris.*

The two members of the Eastover Formation, the lower Claremont Manor and the upper Cobham Bay, contain very different-appearing molluscan assemblages. The Claremont Manor Member, deposited during the late Tortonian cool time, contains a fauna that resembles an impoverished version of the St. Mary's (Windmill Point Member) molluscan fauna. The Cobham Bay, put down during the Messinian warm time, contains a much richer molluscan fauna and more closely resembles assemblages seen in the subsequent Pliocene Yorktown Formation. The Cobham Bay is of special significance in that it contains the last members of several classic Transmarian genera, including gastropods such as the turritellid *Mariacolpus,* the muricid *Mariasalpinx,* the buccinid whelk *Bulliopsis,* the busycon whelk *Coronafulgur,* and the auger shell *Laevihastula,* and the bivalve *Isognomon (Hippochaeta).* Both members of the Eastover Formation are still relatively unexplored paleontologically, and any intensive collecting is bound to yield many new species.

SHELLS OF THE CLAREMONT MANOR MEMBER

Although well-preserved, the gastropod fauna of the Claremont Manor Member comprises only a small number of species. Principal among these are the turritellid *Mariacolpus carinatus* (Gardner, 1948), the naticid moon snail *Euspira* species, the muricid *Ecphora whiteoakensis* Ward and Gilinsky, 1988 (Figure 7.1A,B), and the volute *Volutifusus rappahannockensis* Petuch and Drolshagen, new species (see Systematic Appendix) (Figure 7.1F,G). With the exception of the suspension-feeding *Mariacolpus carinatus,* all the other large gastropods were carnivores, feeding on the abundant bivalve fauna that characterizes the Claremont Manor Member. Of these, several bivalves formed dense, thickly packed beds, including the ark shells *Dallarca clisea* (Dall, 1898) (Figure 7.1D) and *D. rotunda* Ward, 1992 (Figure 7.1E), the small mactrid *Mulinia rappahannockensis* (Gardner, 1944) (Figure 7.4H), the giant pearly oyster *Isognomon (Hippochaeta) sargenti* Petuch and Drolshagen, new species (see Systematic Appendix), the large scallop *Chesapecten*

A
B
C
D
E
F
G
H
I
J
K
L

ceccae Ward, 1992 (Figure 7.1C), and the venerid *Mercenaria druidi gardnerae* Petuch and Drolshagen, new subspecies (see Systematic Appendix) (Figure 7.1K). Living among these dense aggregations were other characteristic Claremont Manor bivalves, including the bittersweet clam *Costaglycymeris virginiae* (Dall, 1898) (Figure 7.1I; first appearance of the genus *Costaglycymeris*), the oyster *Mansfieldostrea brucei* (Ward, 1992) (Figure 7.1J; first appearance of the genus *Mansfieldostrea*), the crassatellid *Marvacrassatella surryensis* (Mansfield, 1929) (Figure 7.1H), and the astartid *Astarte lindae* Petuch and Drolshagen, new species (see Systematic Appendix) (Figure 7.1L).

SHELLS OF THE COBHAM BAY MEMBER

The Messinian Age was relatively warm, and the marine climate within the Albemarle Sea approached warm-temperate to subtropical conditions. This is evidenced by the fauna of the Cobham Bay Member, which contains several warm-water groups that were absent from the Claremont Manor Member. These migrated northward from Florida and the Carolinas and invaded the Rappahannock Subsea by the early Messinian. Some of these include gastropods such as the olive shell *Oliva (Strephona) idonea* Conrad, 1839 (Figure 7.3K) and the cone shell *Gradiconus spenceri* (Ward, 1992) (Figure 7.2G), and bivalves such as the scallop *Carolinapecten urbannaensis* (Mansfield, 1929) (Figure 7.2F), the teredinid *Kuphus fistula* (H.C. Lea, 1843), and the anomiid *Pododesmus philippi* Gardner, 1944 (Figure 7.5J,K). These subtropical Cobham Bay forms coexisted with normally cold-water groups that had apparently adapted to the warmer conditions during the early Messinian. Some of the more important of these include bivalves such as the scallop *Placopecten principoides* (Emmons, 1858) (Figure 7.3C), the astartids *Astarte rappahannockensis* Gardner, 1944 (Figure 7.2J) and *A. cobhamensis* Ward, 1992, the crassatellid *Cyclocardia vautrotorum* Petuch and Drolshagen, new species (see Systematic Appendix) (Figure 7.4C,F), and the moon snail gastropod *Euspira perspectiva* (Rogers and Rogers, 1835) subspecies (Figure 7.4L,M).

FIGURE 7.1 (Facing page) Index fossils for the Claremont Manor Member of the Eastover Formation. A,B = *Ecphora whiteoakensis* Ward and Gilinski, 1988, length 96 mm, lower shell bed of cliff along Urbanna Creek, Urbanna, Middlesex County, Virginia; C = *Chesapecten ceccae* Ward, 1992, length 174 mm, lower shell bed of cliff along Urbanna Creek, Urbanna, Middlesex County, Virginia; D = *Dallarca clisea* (Dall, 1898), length 29 mm, on floor of the Rappahannock River, exposed at extreme low tide, 5 miles south of Bowler's Wharf, Middlesex County, Virginia; E = *Dallarca rotunda* Ward, 1992, length 44 mm, on floor of the Rappahannock River, exposed at extreme low tide, 5 miles south of Bowler's Wharf, Middlesex County, Virginia; F,G = *Volutifusus rappahannockensis* Petuch and Drolshagen, new species (see Systematic Appendix), holotype, length 87 mm, on floor of the Rappahannock River, exposed at extreme low tide, 5 miles south of Bowler's Wharf, Middlesex County, Virginia; H = *Marvacrassatella surryensis* (Mansfield, 1929), length 58 mm, lower shell bed of cliff along Urbanna Creek, Urbanna, Middlesex County, Virginia; I = *Costaglycymeris virginiae* (Dall, 1898), length 55 mm, lower shell bed of cliff along Urbanna Creek, Urbanna, Middlesex County, Virginia; J = *Mansfieldostrea brucei* (Ward, 1992), length 114 mm, lower shell bed of cliff along Urbanna Creek, Urbanna, Middlesex County, Virginia; K = *Mercenaria druidi gardnerae* Petuch and Drolshagen, new subspecies (see Systematic Appendix), holotype, length 68 mm, lower shell bed of cliff along Urbanna Creek, Urbanna, Middlesex County, Virginia; L = *Astarte lindae* Petuch and Drolshagen, new species (see Systematic Appendix), holotype, length 21 mm, lower shell bed of cliff along Urbanna Creek, Urbanna, Middlesex County, Virginia.

A
B
C
D
E
F
G
H
I
J
K
L
M

As in the older Claremont Manor Member, bivalves dominate the ecosystems of the Cobham Bay Member. Shallow-water muddy areas in coastal lagoons, especially those behind the Bayport Archipelago, were dominated by extensive beds of the mactrid marsh clam *Mulinia rappahannockensis* (Gardner, 1944) (Figure 7.4H), the ark shell *Dallarca carolinensis* (Dall, 1898) (Figure 7.2K), and the large venerid *Mercenaria druidi* Ward, 1992 (Figure 7.4J). These muddy coastal lagoons also housed small, scattered beds of other bivalves, mostly the small venerid *Lirophora vredenburgi* Ward, 1992 (Figure 7.2I), the lucinid *Lucinoma contracta* (Say, 1824), and the mactrid *Spisula bowlerensis* (Gardner, 1944). These bivalves were the main prey items for several carnivorous drilling predatory gastropods, including naticid moon snails such as the small *Euspirella permeliae* Petuch and Drolshagen, new species (see Systematic Appendix) (Figure 7.3G,H), the high-spired *Neverita vautrotorum* Petuch and Drolshagen, new species (See Systematic Appendix) (Figure 7.4A,B), the flattened, low-spired *N. taylorae* Petuch and Drolshagen, new species (see Systematic Appendix) (Figure 7.4D,E) and *Euspira* cf. *perspectiva* (Rogers and Rogers, 1835) (Figure 7.4L,M), the small muricid *Urosalpinx rappahannockensis* (Gardner, 1948) (Figure 7.2L,M), and the busycon whelk *Coronafulgur kendrewi* Petuch, 2004 (Figure 7.3L,M; the last *Coronafulgur* species). Small scavenger gastropods such as the buccinid *Bulliopsis bowlerensis* Allmon, 1990 (Figure 7.3D) and the nassariid mud snail *Ilyanassa lindae* Petuch and Drolshagen, new species (see Systematic Appendix) (Figure 7.E,F) were also common on the intertidal mud flats inside the sheltered coastal lagoons.

On the outside of the Bayport Archipelago, in areas with sand and shell hash bottoms and deeper, open oceanic water conditions, large scallops and oysters dominated the Cobham Bay molluscan assemblages. Three large scallops formed the bulk of the pectinid biomass and included *Chesapecten middlesexensis* (Mansfield, 1936) (Figure 7.2C), *Placopecten principoides* (Emmons, 1858) (Figure 7.3C), and *Carolinapecten urbannaensis* (Mansfield, 1929) (Figure 7.2F). The rare, small scallop *Euvola smithi* (Olsson, 1914) also occurred in these large scallop beds. The most obvious large bivalve in the open oceanic areas was the

FIGURE 7.2 (Facing page) Index fossils for the Cobham Bay Member of the Eastover Formation. A,B = *Ecphora kochi* Ward and Gilinski, 1988, length 92 mm, in small Cobham Bay exposure within the Lee Creek Mine, Aurora, Beaufort County, North Carolina; C = *Chesapecten middlesexensis* (Mansfield, 1936), length 158 mm, upper shell bed of cliff along Urbanna Creek, Urbanna, Middlesex County, Virginia; D = *Mansfieldostrea geraldjohnsoni* (Ward, 1992), lower valve, length 121 mm, upper shell bed of cliff along Urbanna Creek, Urbanna, Middlesex County, Virginia; E = *Mansfieldostrea geraldjohnsoni* (Ward, 1992), upper valve, length 98 mm, upper shell bed of cliff along Urbanna Creek, Urbanna, Middlesex County, Virginia; F = *Carolinapecten urbannaensis* (Mansfield, 1929), length 110 mm, Rappahannock River at Urbanna Cliffs, 2 miles south of the mouth of Urbanna Creek, Urbanna, Middlesex County, Virginia; G = *Gradiconus spenceri* (Ward, 1992), length 33 mm, upper shell bed of cliff along Urbanna Creek, Urbanna, Middlesex County, Virginia; H = *Stewartia anodonta* (Say, 1824) subspecies, length 53 mm, upper shell bed in cliff along Urbanna Creek, Urbanna, Middlesex County, Virginia; I = *Lirophora vredenburgi* Ward, 1992, length 26 mm, in cliff, 5 miles south of Bowler's Wharf, Middlesex County, Virginia; J = *Astarte rappahannockensis* Gardner, 1944, length 24 mm, upper shell bed in cliff along Urbanna Creek, Urbanna, Middlesex County, Virginia; K = *Dallarca carolinensis* (Dall, 1898), length 51 mm, at base of cliff on the Rappahannock River, 5 miles south of Bowler's Wharf, Middlesex County, Virginia; L,M = *Urosalpinx rappahannockensis* (Gardner, 1948), length 19 mm, at base of cliff on the Rappahannock River, 5 miles south of Bowler's Wharf, Middlesex County, Virginia.

A
B
C
D
E
F
G
H
I
J
K
L
M

giant pearly oyster *Isognomon (Hippochaeta) sargenti* Petuch and Drolshagen, new species (see Systematic Appendix) (Figure 7.3A,B), which formed immense biohermal, reeflike structures that were scattered across the seafloor. Interspersed among these *Isognomon* bioherms were large individuals of the frilly oyster *Mansfieldostrea geraldjohnsoni* (Ward, 1992) (Figure 7.2D,E).

The oyster bioherms and scallop beds both housed an unusually rich fauna of encrusting epizoonts. Some of these included gastropods such as the vermetid worm shells *Anguinella virginica* (Conrad, 1839) (Figure 7.5M) and *Serpulorbis* cf. *granifera* (Say, 1824) (Figure 7.4K) and bivalves such as the small oyster *Conradostrea greeni* Ward, 1992 and the anomiid *Pododesmus philippi* Gardner, 1944 (Figure 7.5J,K), along with encrusting ahermatypic corals (*Septastrea* species), polychaete worms, and barnacles (Figure 7.5M). The *Isognomon* bioherms also formed the substrate for a large fauna of sessile gastropods including the slipper shells *Crepidula virginica* Conrad, 1871 (Figure 7.5E,F) and *Crepidula (Bostrycapulus) bonnieae* Petuch and Drolshagen, new species (see Systematic Appendix), (Figure 7.5A,B,C; first appearance of the spiny slipper shell subgenus *Bostrycapulus*), and the keyhole limpet *Glyphis lindae* Petuch and Drolshagen, new species (see Systematic Appendix) (Figure 7.5D). The large discinid brachiopod *Discinisca lugubris* (Conrad, 1834) (Figure 7.5L) was abundant in these pearly oyster bioherms, and upper valves are commonly collected on beaches all along Urbanna Creek and the Rappahannock River near Urbanna, Middlesex County, Virginia.

Scattered between the oyster bioherms and scallop beds, on clean sand and shell hash bottoms, were large aggregations of smaller bivalves. Some of the more abundant species included the bittersweet clam *Costaglycymeris mixoni* Ward, 1992 (Figure 7.4I), the noetiid ark shell *Striarca centenaria* (Say, 1824) (Figure 7.3J), the cockle shell *Chesacardium blountense* (Mansfield, 1932) (also found in the Messinian Miocene of the Florida Panhandle), the crassatellids *Marvacrassatella urbannaensis* (Mansfield, 1929) (Figure 7.3I) and *Cyclocardia vautrotorum* Petuch and Drolshagen, new species (Figure 7.4C,F), the lucinid *Stewartia anodonta* (Say, 1824) subspecies (Figure 7.2H), the astartid *Astarte rappahannockensis* Gardner, 1944 (Figure 7.2J), and the venerids *Dosinia blountana* (Mansfield, 1932)

FIGURE 7.3 (Facing page) Index fossils for the Cobham Bay Member of the Eastover Formation. A,B = *Isognomon (Hippochaeta) sargenti* Petuch and Drolshagen, new species (see Systematic Appendix), holotype, length (left valve) 242 mm, in cliff along Urbanna Creek, Urbanna, Middlesex County, Virginia; C = *Placopecten principoides* (Emmons, 1858), length 135 mm, Cobham Bay beds at Murfreesboro, Hertford County, North Carolina; D = *Bulliopsis bowlerensis* Allmon, 1990, length 24 mm, at base of cliff on the Rappahannock River, 5 miles south of Bowler's Wharf, Middlesex County, Virginia; E,F = *Ilyanassa lindae* Petuch and Drolshagen, new species (see Systematic Appendix), holotype, length 14 mm, at base of cliff on the Rappahannock River, 5 miles south of Bowler's Wharf, Middlesex County, Virginia; G,H = *Euspirella permeliae* Petuch and Drolshagen, new species (see Systematic Appendix), holotype, length 15 mm, at base of cliff on the Rappahannock River, 5 miles south of Bowler's Wharf, Middlesex County, Virginia; I = *Marvacrassatella urbannaensis* (Mansfield, 1929), length 54 mm, upper shell bed in cliff along Urbanna Creek, Urbanna, Middlesex County, Virginia; J = *Striarca centenaria* (Say, 1824), length 29 mm, upper shell bed in cliff along Urbanna Creek, Urbanna, Middlesex County, Virginia; K = *Oliva (Strephona) idonea* Conrad, 1839, length 25 mm, in small Cobham Bay exposure within the Lee Creek Mine, Aurora, Beaufort County, North Carolina; L,M = *Coronafulgur kendrewi* Petuch, 2004, length 122 mm, at base of cliff on the Rappahannock River, 5 miles south of Bowler's Wharf, Middlesex County, Virginia.

A
B
C
D
E
F
G
H
I
J
K
L
M

(Figure 7.5I; also found in the Messinian Miocene of the Florida Panhandle), *Mercenaria druidi* Ward, 1992 (Figure 7.4J), *Macrocallista* species, and *Lirophora vredenburgi* Ward, 1992 (Figure 7.2I). Small, scattered beds of the turritellid gastropods *Torcula roperorum* Petuch and Drolshagen, new species (See Systematic Appendix) (Figure 7.5G,H) and *Mariacolpus plebeia* (Say, 1824) (Figure 7.4G; last-living species of the genus *Mariacolpus*) also occurred on these clean sand seafloors. These bivalves and turritellid gastropods were the principal prey items for the drilling gastropods *Ecphora kochi* Ward and Gilinsky, 1988 (Figure 7.2A,B; the only Cobham Bay ecphora) and the previously mentioned moon snails *Neverita vautrotorum, N. taylorae,* and *Euspira perspectiva.*

SYSTEMATIC LIST OF EASTOVER MOLLUSKS

The following is a list of the more important index fossils of the Eastover Formation, arranged systematically and by member.

EASTOVER FORMATION

Claremont Manor Member

Gastropoda

Turritellidae

Mariacolpus carinatus (Gardner, 1948)

Naticidae

Euspira species (*heros* complex)

Euspirella species (close to *E. permeliae*)

Neverita species (close to *N. asheri*)

FIGURE 7.4 **(Facing page)** Index fossils for the Cobham Bay Member of the Eastover Formation. A,B = *Neverita vautrotorum* Petuch and Drolshagen, new species (see Systematic Appendix), holotype, length 39 mm, upper shell bed in cliff along Urbanna Creek, Urbanna, Middlesex County, Virginia; C = *Cyclocardia vautrotorum* Petuch and Drolshagen, new species (see Systematic Appendix), holotype, length 25 mm, upper shell bed in cliff along Urbanna Creek, Urbanna, Middlesex County, Virginia; D,E = *Neverita taylorae* Petuch and Drolshagen, new species (see Systematic Appendix), holotype, length 25 mm, upper shell bed in cliff along Urbanna Creek, Urbanna, Middlesex County, Virginia; F = *Cyclocardia vautrotorum* Petuch and Drolshagen, new species, 16 mm juvenile specimen showing heavy cords, upper shell bed in cliff along Urbanna Creek, Urbanna, Middlesex County, Virginia; G = *Mariacolpus plebeia* (Say, 1824), length 51 mm, at base of cliff on the Rappahannock River, 5 miles south of Bowler's Wharf, Middlesex County, Virginia; H = *Mulinia rappahannockensis* (Gardner, 1944), length 24 mm, at base of cliff on the Rappahannock River, 5 miles south of Bowler's Wharf, Middlesex County, Virginia; I = *Costaglycymeris mixoni* Ward, 1992, length 35 mm, upper shell bed in cliff along Urbanna Creek, Urbanna, Middlesex County, Virginia; J = *Mercenaria druidi* Ward, 1992, length 82 mm, at base of cliff on the Rappahannock River, 5 miles south of Bowler's Wharf, Middlesex County, Virginia; K = *Serpulorbis* cf. *granifera* (Say, 1824), length 37 mm, growing on the back of the oyster *Mansfieldostrea geraldjohnsoni;* L,M = *Euspira perspectiva* (Rogers and Rogers, 1835), length 28 mm, upper shell bed in cliff along Urbanna Creek, Urbanna, Middlesex County, Virginia.

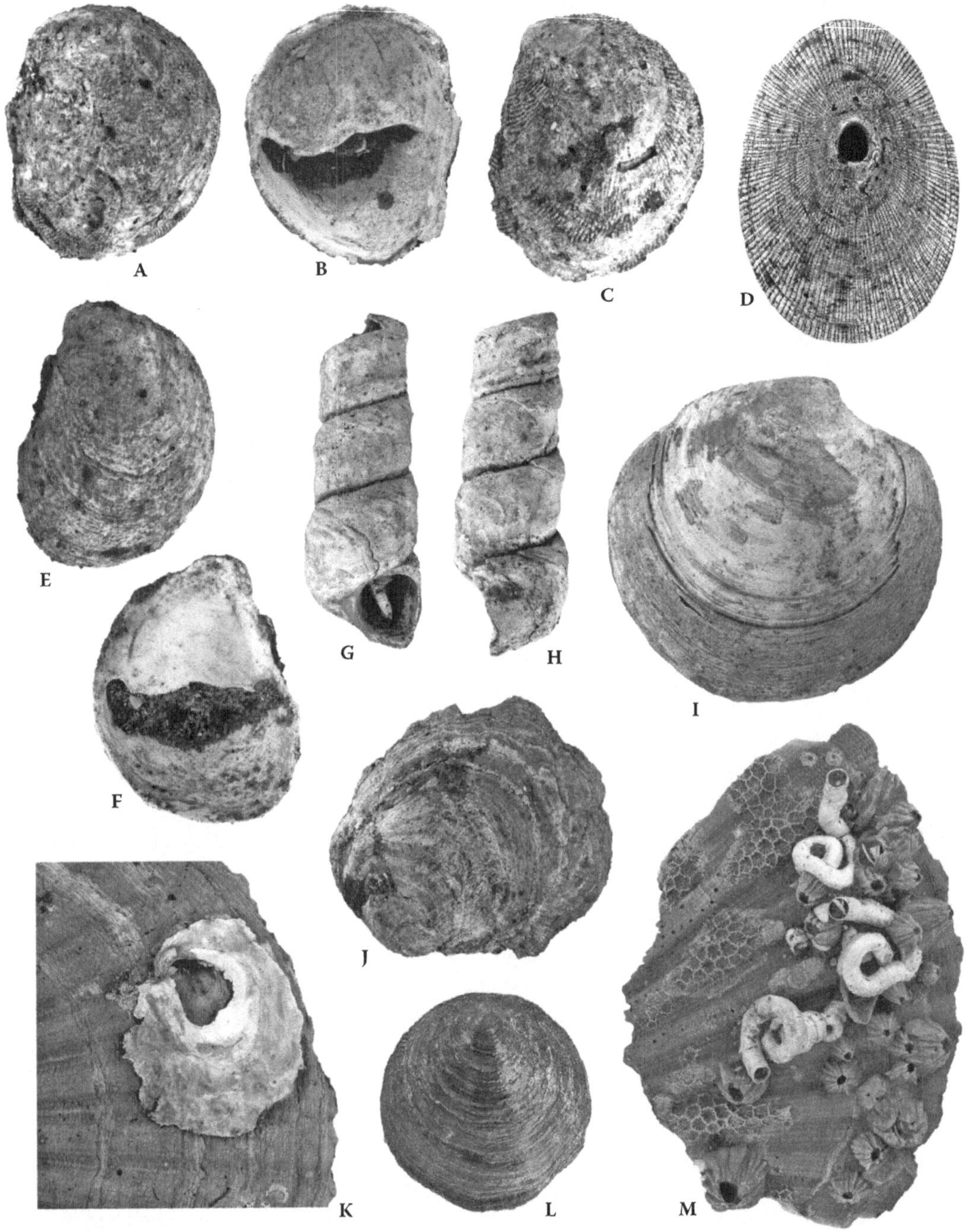
A
B
C
D
E
F
G
H
I
J
K
L
M

Muricidae
 Ecphora whiteoakensis Ward and Gilinsky, 1988 (Figure 7.1A,B)
Volutidae
 Volutifusus rappahannockensis Petuch and Drolshagen, new species (Figure 7.1F,G)
Bivalvia
Arcidae
 Dallarca clisea (Dall, 1898) (Figure 7.1D)
 Dallarca rotunda Ward, 1992 (Figure 7.1E)
 Dallarca virginiae (Dall, 1898)
Glycymeridae
 Costaglycymeris virginiae (Dall, 1898) (Figure 7.1I)
Pectinidae
 Chesapecten ceccae Ward, 1992 (Figure 7.1C)
Ostreidae
 Mansfieldostrea brucei (Ward, 1992) (Figure 7.1J)
Isognomonidae
 Isognomon (Hippochaeta) sargenti Petuch and Drolshagen, new species
Lucinidae
 Lucinoma contracta subspecies
 Stewartia anodonta (Say, 1824)
Mactridae
 Mulinia rappahannockensis (Gardner, 1944)
Veneridae
 Lirophora dalli (Olsson, 1914)
 Mercenaria druidi gardnerae Petuch and Drolshagen, new subspecies (Figure 7.1K)
Psammobiidae
 Pleiorhytis species
Astartidae
 Astarte lindae Petuch and Drolshagen, new species (Figure 7.1L)
Crassatellidae
 Marvacrassatella surryensis (Mansfield, 1929) (Figure, 7.1H)

FIGURE 7.5 **(Facing page)** Index fossils for the Cobham Bay Member of the Eastover Formation. A,B = *Crepidula (Bostrycapulus) bonnieae* Petuch and Drolshagen, new species (see Systematic Appendix), holotype, length 29 mm; C = *Crepidula (Bostrycapulus) bonnieae* Petuch and Drolshagen, new species, paratype, length 23 mm, showing fine ribbed sculpture on dorsum; D = *Glyphis lindae* Petuch and Drolshagen, new species (see Systematic Appendix), holotype, length 43 mm; E,F = *Crepidula virginica* Conrad, 1871, length 22 mm; G,H = *Torcula roperorum* Petuch and Drolshagen, new species (see Systematic Appendix), holotype (incomplete), length 48 mm; I = *Dosinia blountana* Mansfield, 1932, length 82 mm; J = *Pododesmus philippi* Gardner, 1944, length (upper valve) 41 mm; K = *Pododesmus philippi* Gardner, 1944, length (lower valve), 38 mm, attached to *Chesapecten middlesexensis;* L = the brachiopod *Discinisca lugubris* (Conrad, 1834), length 28 mm; M = *Anguinella virginica* (Conrad, 1839), largest individual 22 mm, three individuals attached to fragment of *Chesapecten middlesexensis;* the coral *Septastrea* and small barnacles grew together with the *Anguinella* worm shells. All specimens were collected in the upper shell bed in a cliff along Urbanna Creek, Urbanna, Middlesex County, Virginia.

Euloxidae
Euloxa latisulcata (Conrad, 1839)
Glossidae
Glossus fraterna (Say, 1824)
Hiatellidae
Panopea goldfussi Wagner, 1839

Cobham Bay Member
Gastropoda
Fissurellidae
Glyphis lindae Petuch and Drolshagen, new species (Figure 7.5D)
Crepidulidae
Crepidula virginica Conrad, 1871 (Figure 7.5E,F)
Crepidula (Bostrycapulus) bonnieae Petuch and Drolshagen, new species (Figure 7.5A,B,C)
Turritellidae
Mariacolpus plebeia (Say, 1824) (Figure 7.4G)
Torcula roperorum Petuch and Drolshagen, new species (Figure 7.5G,H)
Vermetidae
Anguinella virginica (Conrad, 1839) (Figure 7.5M)
Serpulorbis cf. *granifera* (Say, 1824) (Figure 7.4K)
Naticidae
Euspira cf. *perspectiva* (Rogers and Rogers, 1835) (Figure 7.4L,M)
Euspirella permeliae Petuch and Drolshagen, new species (Figure 7.3G,H)
Neverita taylorae Petuch and Drolshagen, new species (Figure, 7.4D,E)
Neverita vautrotorum Petuch and Drolshagen, new species (Figure 7.4A,B)
Muricidae
Ecphora kochi Ward and Gilinsky, 1988 (Figure 7.2A,B)
Mariasalpinx species (*M. vokesae* complex)
Urosalpinx rappahannockensis (Gardner, 1948) (Figure 7.2L,M)
Busyconidae
Busycotypus cf. *alveatum* (Conrad, 1863)
Coronafulgur kendrewi Petuch, 2004 (Figure 7.3L,M)
Buccinidae
Bulliopsis bowlerensis Allmon, 1990 (Figure 7.3D)
Nassariidae
Ilyanassa lindae Petuch and Drolshagen, new species (Figure 7.3E,F)
Olividae
Oliva (Strephona) idonea Conrad, 1839 (Figure 7.3K)
Conidae
Gradiconus spenceri (Ward, 1992) (Figure 7.2G)
Terebridae
Laevihastula cf. *simplex* (Conrad, 1830)

Bivalvia

Arcidae

Dallarca carolinensis (Dall, 1898) (Figure 7.2K)

Rasia arata (Say, 1824)

Noetiidae

Striarca centenaria (Say, 1824) (Figure 7.3J)

Glycymeridae

Costaglycymeris mixoni Ward, 1992 (Figure 7.4I)

Pectinidae

Carolinapecten urbannaensis (Mansfield, 1929) (Figure 7.2F)

Chesapecten middlesexensis (Mansfield, 1936) (Figure 7.2C)

Euvola smithi (Olsson, 1914)

Placopecten principoides (Emmons, 1858) (Figure 7.3C)

Ostreidae

Conradostrea greeni Ward, 1992

Mansfieldostrea geraldjohnsoni (Ward, 1992) (Figure 7.2D,E)

Lucinidae

Lucinoma contracta (Say, 1824) subspecies

Stewartia anodonta (Say, 1824) subspecies (Figure 7.2H)

Isognomonidae

Isognomon (Hippochaeta) sargenti Petuch and Drolshagen, new species (Figure 7.3A,B)

Anomiidae

Pododesmus philippi Gardner, 1944 (Figure 7.5J,K)

Chamidae

Chama cf. *corticosa* (Conrad, 1833)

Mactridae

Mulinia rappahannockensis (Gardner, 1944) (Figure 7.4H)

Spisula bowlerensis (Gardner, 1944)

Veneridae

Dosinia blountana Mansfield, 1932 (Figure 7.5I)

Lirophora vredenburgi Ward, 1992 (Figure 7.2I)

Macrocallista species

Mercenaria druidi Ward, 1992 (Figure 7.4J)

Crassatellidae

Cyclocardia vautrotorum Petuch and Drolshagen, new species (Figure 7.4C,F)

Marvacrassatella urbannaensis (Mansfield, 1929) (Figure 7.3I)

Teredinidae

Kuphus fistula (H.C. Lea, 1843)

Psammobiidae

Pleiorhytis species

Cardiidae

Chesacardium blountense (Mansfield, 1932)

Astartidae
Astarte cobhamensis Ward, 1992
Astarte rappahannockensis Gardner, 1944 (Figure 7.2J)
Euloxidae
Euloxa latisulcata (Conrad, 1839)
Glossidae
Glossus fraterna (Say, 1824)
Hiatellidae
Panopea goldfussi Wagner, 1839

References

Allmon, W.D. 1990. Review of the *Bullia* Group (Gastropoda: Nassariidae) with comments on its evolution, biogeography, and phylogeny. *Bulletins of American Paleontology* 99 (335): 1–179.

Briggs, J.C. 1974. *Marine Zoogeography.* McGraw-Hill Book Company, New York, 270 pp.

Campbell, L.D. 1993. *Pliocene Molluscs from the Yorktown and Chowan River Formations in Virginia.* Virginia Division of Mineral Resources, Publication 127. Department of Mines, Minerals, and Energy, Charlottesville, VA, 259 pp.

Clark, W.B. 1904. Introduction and general stratigraphic relations. In *The Miocene Deposits of Maryland,* W.B. Clark, G.B. Shattuck, and W.H. Dall, Editors. Department of Natural Resources, Maryland Geological Survey, Baltimore, MD, pp. 23–64.

Conrad, T.A. 1830. On the geology and organic remains of a part of the Peninsula of Maryland. *Journal of the Academy of Natural Sciences of Philadelphia* 6 (2): 205–230, plates 9, 10.

Conrad, T.A. 1832. *Fossil Shells of the Tertiary Formations of North America* Volume 1. Judah Dobson, Publisher, Philadelphia, PA, pp.1–20, plates 9,10.

Conrad, T.A. 1833. On some new fossil and recent shells of the United States. *American Journal of Science and Arts* 23 (17): 339–346.

Conrad, T.A. 1838. *Fossils of the Tertiary Formations of the United States.* Judah Dobson, Publisher, Philadelphia, PA, 32 pp., 18 plates.

Conrad, T.A. 1840. *Fossils of the Medial Tertiary of the United States* No. 2. Judah Dobson, Publisher, Philadelphia, PA, pp. 33–56, plates 18–29.

Conrad, T.A. 1841. Twenty-six new species of fossil shells, Medial Tertiary deposits of Calvert Cliffs, Maryland. *Proceedings of the Academy of Natural Sciences of Philadelphia* 1 (3): 28–33.

Conrad, T.A. 1842. Description of twenty-four new species of fossil shells, chiefly from the Tertiary deposits of Calvert Cliffs. *Journal of the Academy of Natural Sciences of Philadelphia* 8 (2): 183–190.

Conrad, T.A. 1843. Descriptions of a new genus, and of twenty-nine new Miocene, and one new Eocene fossil shells of the United States. *Proceedings of the Academy of Natural Sciences of Philadelphia* 1 (3): 305–311.

Conrad, T.A. 1863. Descriptions of new genera, subgenera, and species of Tertiary and recent shells. *Proceedings of the Academy of Natural Sciences of Philadelphia* 1862: 284–291.

Conrad, T.A. 1869. Descriptions of and references to Miocene shells of the Atlantic Slope and descriptions of two new supposed Cretaceous species. *American Journal of Conchology* 4 (4): 278–279, plates 19, 20.

Dall, W.H. 1892. Contributions to the Tertiary Fauna of Florida with especial reference to the Miocene Silex-beds of Tampa and the Pliocene beds of the Caloosahatchie River. *Transactions of the Wagner Free Institute of Science of Philadelphia* 3 (1): 1–200.

Dall, W.H. 1898. Contributions to the Tertiary fauna of Florida. *Transactions of the Wagner Free Institute of Science of Philadelphia* 3 (4): 571–947.

Emmons, E. 1858. *Agriculture of the Eastern Counties, Together with Descriptions of the Fossils of the Marl Beds.* Report of the North Carolina Geological Survey, Henry D. Turner, Publisher, Raleigh, NC.

Gardner, J.A. 1944. *Mollusca from the Miocene and Lower Pliocene of Virginia and North Carolina.* Part 1. Pelecypoda. U.S. Geological Survey Professional Paper 199-A, pp.1–178, plates 1–23.

Gardner, J.A. 1948. *Mollusca from the Miocene and Lower Pliocene of Virginia and North Carolina.* Part 2. Scaphopoda and Gastropoda. U.S. Geological Survey Professional Paper 199-B, pp. 179–310, plates 24–38.

Gernant, R.E. 1970. *Paleoecology of the Choptank Formation (Miocene) of Maryland and Virginia.* Maryland Geological Survey Report of Investigation No. 12, 90 pp.

Gibson, T.G. 1962. Revision of the Turridae of the Miocene St. Mary's Formation of Maryland. *Journal of Paleontology* 36 (2): 225–246.

Gibson, T.G. 1983. Stratigraphy of Miocene through lower Pleistocene strata of the United States Central Atlantic Coastal Plain. In *Geology and Paleontology of the Lee Creek Mine, North Carolina.* 1. Smithsonian Contributions to Paleobiology, ed. C.E. Ray, Smithsonian Institution Press, Washington, DC, No. 53, pp. 355–453.

Gibson, T.G. 1987. Miocene and Pliocene Pectinidae (Bivalvia) from the Lee Creek Mine and adjacent areas. In *Geology and Paleontology of the Lee Creek Mine, North Carolina* 2. Smithsonian Contributions to Paleobiology, ed. C.E. Ray, Smithsonian Institution Press, Washington, DC, No. 61, pp. 31–112.

Glenn, L.C. 1904. Mollusca, Pelecypoda. In *Systematic Paleontology of the Miocene Deposits of Maryland.* Miocene Volume, Department of Natural Resources, Maryland Geological Survey, Baltimore, MD, pp. 274–401, plates 65–108.

Green, J. 1830. Monograph of the cones of North America, including three new species. *Transactions of the Albany Institute* 1 (7): 121–125.

Mansfield, W.C. 1929. New fossil mollusks from the Miocene of Virginia and North Carolina, with a brief outline of the divisions of the Chesapeake Group. *Proceedings of the United States National Museum* 74 (14; 2759): 1–11.

Mansfield, W.C. 1936. Stratigraphic significance of Miocene, Pliocene, and Pleistocene Pectinidae in the southeastern United States. *United States Journal of Paleontology* 10 (3): 168–192.

Martin, G.C. 1904. Mollusca, Gastropoda. In *Systematic Paleontology of the Miocene Deposits of Maryland.* Miocene Volume, Department of Natural Resources, Maryland Geological Survey, Baltimore, MD, pp. 131–270, plates 39–63.

Oleksyshyn, J. 1959. Some new species of Miocene Mollusca from Maryland. *Journal of Paleontology* 33 (1): 29–32.

Oleksyshyn, J. 1960. Some new species of Miocene Mollusca from Maryland and Virginia. *Proceedings of the Pennsylvania Academy of Science* 34: 101–106.

Petuch, E.J. 1988a. New species of *Ecphora* and Ecphorine thaidids from the Miocene of Chesapeake Bay, Maryland. *Bulletin of Paleomalacology* 1 (1): 1–16 (copies available at the National Museum of Natural History, Smithsonian Institution).

Petuch, E.J. 1988b. New species of *Trisecphora* from the Miocene of Maryland, U.S.A. *Bulletin of Paleomalacology* 1 (2): 36–39.

Petuch, E.J. 1988c. New gastropods from the Maryland Miocene. *Bulletin of Paleomalacology* 1 (4): 69–80.

Petuch, E.J. 1988d. *Neogene History of Tropical American Mollusks: Biogeography and Evolutionary Patterns of Tropical Western Atlantic Mollusca.* The Coastal Education and Research Foundation, Charlottesville, VA.

Petuch, E.J. 1989. *Field Guide to the Ecphoras.* The Coastal Education and Research Foundation, Charlottesville, VA, 140 pp. (15 February).

Petuch, E.J. 1992. New ecphoras (Gastropoda: Thaididae: Ecphorinae) from the Calvert Formation of Maryland (Langhian Miocene). *The Nautilus* 106 (2): 68–71.

Petuch, E.J. 1993. Patterns of diversity and extinction in Transmarian Muricacean, Buccinacean, and Conacean gastropods. *The Nautilus* 106 (4): 155–173.

Petuch, E.J. 1997. *Coastal Paleoceanography of Eastern North America (Miocene–Pleistocene).* Kendall/Hunt Publishing Company, Dubuque, IA, 373 pp.

Petuch, E.J. 2004. *Cenozoic Seas: The View from Eastern North America.* CRC Press, New York and London, 308 pp.

Petuch, E.J. 2008. *The Geology of the Florida Keys and Everglades: An Illustrated Field Guide to Florida's Hidden Beauty.* Thomson/Cengage Publishers, Mason, OH, 84 pp.

Petuch, E.J. and C.E. Roberts. 2007. *The Geology of the Everglades and Adjacent Areas.* CRC Press, New York and London, 212 pp.

Poag, C.W. 1999. *Chesapeake Invader: Discovering America's Giant Meteor Crater.* Princeton University Press, Princeton, NJ, 183 pp.

Poag, C.W., C. Koeberl, and W.U. Reimold. 2004. *The Chesapeake Bay Crater: Geology and Geophysics of a Late Eocene Submarine Impact Structure.* Springer-Verlag Publishers, New York, 522 pp.

Say, T. 1824. An account of some of the fossil shells of Maryland. *Journal of the Academy of Natural Sciences of Philadelphia* 4 (1): 124–155.

Schoonover, L.M. 1941. A stratigraphic study of the mollusks of the Calvert and Choptank Formations of southern Maryland. *Bulletins of American Paleontology* 25 (94B): 1–134.

Shattuck, G.B. 1904. Geological and paleontological relations, with a review of earlier investigations. In *The Miocene Deposits of Maryland,* ed. W.B. Clark, G.B. Shattuck, and W.H. Dall, Department of Natural Resources, Maryland Geological Survey, Baltimore, MD, pp. 33–137.

Tucker, H.I. 1934. Some Atlantic Coast Tertiary Pectinidae. *The American Midland Naturalist* 15 (5): 612–621.

Tucker, H.I. 1936. The Atlantic and Gulf Coast Tertiary Pectinidae of the United States. *The American Midland Naturalist* 17 (2): 471–490.

Valentine, J.W. 1973. *Evolutionary Paleoecology of the Marine Biosphere.* Prentice-Hall, Inc. Publishers, Englewood, NJ, 511 pp.

Wagner, W. 1839. Description of five new fossils, of the Older Pliocene Formation of Maryland and North Carolina. *Journal of the Academy of Natural Sciences of Philadelphia* 8 (1): 51–53.

Ward, L.W. 1985. Late Tertiary evolution of Salisbury Embayment. *American Association of Petroleum Geologists* 69 (9): 1451.

Ward, L.W. 1992. *Molluscan Biostratigraphy of the Miocene, Middle Atlantic Coastal Plain of North America.* Virginia Museum of Natural History Memoir 2, Martinsburg, VA, 159 pp., 26 plates.

Ward, L.W. 1998. Mollusks from the lower Miocene Pollack Farm site, Kent County, Delaware: A preliminary analysis. In *Geology And Paleontology of The Lower Miocene Pollack Farm Site, Delaware,* ed. R.N. Benson, Special Publication 21, Delaware Geological Survey, pp. 59–131.

Ward, L.W. and B.W. Blackwelder. 1975. *Chesapecten, a New Genus of Pectinidae (Mollusca: Bivalvia) from the Miocene and Pliocene of Eastern North America.* U.S. Geological Survey Professional Paper 861, 24 pp.

Ward, L.W. and B.W. Blackwelder. 1980. *Stratigraphic Revision of Upper Miocene and Lower Pliocene Beds of the Chesapeake Group—Middle Atlantic Coastal Plain.* U.S. Geological Survey Bulletin 1482-D, 61 pp.

Ward, L.W. and N.F. Gilinsky. 1988. *Ecphora* (Gastropoda: Muricidae) from the Chesapeake Group of Maryland and Virginia. *Academy of Natural Sciences of Philadelphia, Notulae Naturae* 459:1–21.

Whitfield, R.P. 1894. *Mollusca and Crustacea of the Miocene Formations of New Jersey.* U.S. Geological Survey Monograph 24, 195 pp., 24 plates.

The giant razor-toothed crocodilian *Thecachampsa sicaria* Cope, 1869, an inhabitant of the estuaries and mud flats of the St. Mary's Subsea.

Systematic Appendix
Descriptions of New Species, Subspecies, and Genera

While undertaking the research for this book, we found that 28 important stratigraphic index fossils and two genera were unnamed. To address this problem, we here describe these new Miocene taxa. The holotypes of the new species are deposited in the molluscan paleontological collections of the Paleontological Research Institute, Ithaca, New York and bear PRI numbers. These include:

NEW SPECIES AND SUBSPECIES

Gastropoda

Glyphis lindae new species, Eastover Formation, PRI 9907
Crepidula (Bostrycapulus) bonnieae new species, Eastover Formation, PRI 9908
Crepidula (Bostrycapulus) bonnieae new species, paratype, PRI 9909
Xenophora lindae new species, Calvert Formation, PRI 9910
Mariacolpus prunicola new species, Calvert Formation, PRI 9911
Torcula roperorum new species, Eastover Formation, PRI 9912
Euspirella chancellorensis new species, St. Mary's Formation, PRI 9913
Euspirella permeliae new species, Eastover Formation, PRI 9914
Euspirella santamaria new species, St. Mary's Formation, PRI 9915
Neverita taylorae new species, Eastover Formation, PRI 9916
Neverita vautrotorum new species, Eastover Formation, PRI 9917
Ecphora amyae new species, Choptank Formation, PRI 9918
Patuxentrophon plummerae new species, Calvert Formation, PRI 9919
Trisecphora bartoni new species, Calvert Formation, PRI 9920
Urosalpinx subrusticus bartoni new subspecies, St. Mary's Formation, PRI 9921
Busycotypus martini new species, St. Mary's Formation, PRI 9922
Ilyanassa chancellorensis new species, St. Mary's Formation, PRI 9923
Ilyanassa lindae new species, Eastover Formation, PRI 9924
Volutifusus rappahannockensis new species, Eastover Formation, PRI 9925
Marianarona gernanti new species, Choptank Formation, PRI 9926
Marianarona lindae new species, St. Mary's Formation, PRI 9927

Bivalvia

Dallarca chancellorensis new species, St. Mary's Formation, PRI 9928
Dallarca covepointensis new species, St. Mary's Formation, PRI 9929
Isognomon (Hippochaeta) sargenti new species, Eastover Formation, PRI 9930
Astarte lindae new species, Eastover Formation, PRI 9931

Cyclocardia calvertensis new species, Calvert Formation, PRI 9932
Cyclocardia vautrotorum new species, Eastover Formation, PRI 9933
Mercenaria druidi gardnerae new subspecies, Eastover Formation, PRI 9934
Mercenaria lindae new species, St. Mary's Formation, PRI 9935

NEW GENERA

Gastropoda
Euspirella new genus (type species: *Euspirella chancellorensis* Petuch and Drolshagen, new species)

Bivalvia
Mansfieldostrea new genus (type species: *Ostrea compressirostra* Say, 1824)

Gastropoda
Prosobranchia
Vetigastropoda
Fissurelloidea
Fissurellidae
Diodorinae
Genus *Glyphis* Carpenter, 1856
Glyphis lindae new species
(Figure 7.5D)

Description: Shell of average size for genus, broadly oval in shape and slightly laterally compressed; flattened in profile; apical foramen proportionally large, oval in shape, bounded by large thick callus on interior; apical foramen concave in profile, being highest at anterior and posterior ends; shell sculptured with approximately 160 fine radiating ribs; radiating ribs crossed by fine concentric ribs, producing beaded appearance; rib pattern generally consisting of single large rib bounded by two thinner, smaller ribs; edge of interior sculpted with numerous crenulations, with each corresponding to an exterior rib.

Type Material: Holotype PRI 9907, length 43 mm, width 30 mm.

Type Locality: Eastern side of Urbanna Creek, Urbanna, Middlesex County, Virginia, 2 miles from the State Road 603 bridge, at base of low cliff.

Stratigraphic Range: Confined to the Cobham Bay Member of the Eastover Formation, Messinian Miocene of Virginia.

Etymology: Named for Linda J. Petuch, wife of the senior author.

Discussion: The new species is most similar to *Glyphis fluviamariana* (Petuch, 1988) (Figure 6.4F) from the Windmill Point Member of the St. Mary's Formation, but differs in being a larger shell with coarser ribbed sculpture and in having a more rounded apical foramen. In *Glyphis lindae,* the apical foramen is posterior to the center of the shell, while in *G. fluviamariana,* the apical foramen is more centrally located. The new species is also similar to *Glyphis redimicula* (Say, 1824) from the Sunken Meadow and Rushmere Members of the Pliocene Yorktown Formation, but differs in being a more flattened shell with finer ribbed sculpture and in having the apical foramen more posteriorly located.

Glyphis lindae represents the perfect morphological intermediate between the St. Mary's (Tortonian Miocene) *G. fluviamariana* and the Yorktown (Zanclean and Piacenzian Pliocene) *G. redimicula.*

Caenogastropoda
Crepiduloidea
Crepidulidae
Crepidulinae
Genus *Crepidula* Lamarck, 1799
Subgenus *Bostycapulus* Olsson and Harbison, 1953
Crepidula (Bostrycapulus) bonnieae new species
(Figure 7.5A,B,C)

Description: Shell of average size for subgenus, oval and rounded in outline and somewhat flattened; apex distinctly recurved, flattened; apertural shelf edge distinctly sinusoidal, with slightly raised central keel; shelf covering fully ½ of entire apertural area; sculpture on dorsum composed of extremely numerous, fine, closely packed vermiform riblets that radiate out from shell apex; riblets often bifurcate, producing fingerprint-type pattern; sculpture on dorsum of holotype obliterated by stacking of other individuals.

Type Material: Holotype, PRI 9908, length 29 mm, width 26 mm; paratype (with better dorsal sculpture), PRI 9909, length 23 mm, width 20 mm.

Type Locality: Eastern side of Urbanna Creek, Urbanna, Middlesex County, Virginia, 2 miles from the State Road 603 bridge, at base of low cliff.

Stratigraphic Range: Confined to the Cobham Bay Member of the Eastover Formation, Messinian Miocene of Virginia.

Etymology: Named for Mrs. Bonnie Vautrot of Urbanna, Virginia.

Discussion: The new species is the oldest-known representative of the subgenus *Bostrycapulus* (the spiny slipper shells). As the oldest and most primitive form, the new species lacks distinct spines, having instead numerous very fine vermiform riblets. Of the known *Bostrycapulus* species, *Crepidula bonnieae* is most similar to *C. (Bostrycapulus) costata* Morton, 1829 from the Pliocene Yorktown (Virginia), Duplin (Carolinas), and Tamiami (Florida) Formations. The new species differs from its Pliocene descendant in having much finer, more numerous, and more closely packed radiating riblets and in lacking any knobs or spines.

Xenophoridae
Genus *Xenophora* G. Fischer, 1807
Xenophora lindae new species
(Figure 4.4N,O)

Description: Shell small for genus, trochoidal, with flattened margins; spire flattened, broadly scalariform; dorsum ornamented with regularly spaced attached shells, with holotype having preferred *Astarte* and *Cyclocardia* shells; areas between attached shells with rough, slightly pebbled texture, often with low rib-like crenulations; base of shell flattened, smooth, ornamented with numerous very faint, slightly incised spiral threads; aperture compressed, slit-like.

Type Material: Holotype, PRI 9910, length 31 mm, width 28 mm.

Type Locality: In Shattuck Zone 10, Plum Point, Calvert County, Maryland.

Stratigraphic Range: Known only from Shattuck Zone 10 of the Plum Point Member, Calvert Formation, Langhian Miocene of Maryland.

Etymology: Named for Linda J. Petuch, wife of the senior author.

Discussion: The new species is most similar to *Xenophora delecta* (Guppy, 1876) from the Burdigalian Miocene of the Dominican Republic (Baitoa Formation), Florida (Chipola Formation), Panama (lower Gatun Formation), and Venezuela (Cantaure Formation). *Xenophora lindae* differs from the older, widespread *X. delecta* in being a smaller, lower-spired shell with a much smoother, less-sculptured shell base. The base of *X. delecta* is sculptured with tiny, strong beads, while the base of *X. lindae* has only faint spiral lines that follow the coiling of the shell. The new species was originally illustrated by Martin (1904: plate 59, figures 7a,7b,7c), but was misidentified as "*Xenophora conchyliophora* Born" (a living species from the Recent tropical western Atlantic). *Xenophora lindae* is the only species of carrier shell ever found in the Chesapeake Miocene.

Cerithioidea
Turritellidae
Turritellinae
Genus *Mariacolpus* Petuch, 1988
Mariacolpus prunicola new species
(Figure 4.4L)

Description: Shell small for genus, narrow, with distinctly flattened, straight-sided whorls; suture strongly indented, producing overhanging edges on previous whorls; edges of overhanging whorls with single large, carina-like cord; whorls ornamented with 10 large, well-developed, equal-sized spiral cords; base of shell proportionally broad; aperture roughly rectangular in shape.

Type Material: Holotype, PRI 9911, length 27 mm, width 8 mm; two specimens in the research collection of the senior author.

Type Locality: In Shattuck Zone 10, Plum Point, Calvert County, Maryland.

Stratigraphic Range: Confined to Shattuck Zone 10 of the Plum Point Member, Calvert Formation, Langhian Miocene of Maryland.

Etymology: Named for Plum Point, Maryland, the type locality.

Discussion: Although found along with its congener *Mariacolpus aequistriata* (Conrad, 1863) in Shattuck Zone 10, *M. prunicola* is never as common as *M. aequistriata.* The new species differs from its congener in being a smaller, stockier shell with distinctly flattened, straight sides on its whorls, not rounded like those of *M. aequistriata* (Figure 4.4K). Because of its flattened, sloping whorls, the aperture of *M. prunicola* is roughly rectangular in shape while the aperture of *M. aequistriata,* with its rounded whorls, is distinctly oval. The new species also is a more coarsely sculptured shell, with 10 large spiral cords, while *M. aequistriata* is a more delicately sculptured species, with 15 to 17 very fine, threadlike cords. *Mariacolpus prunicola,* with its straight sides and coarse sculpture, appears to be the ancestor of the coarse-sculptured (8 cords) *M. octonaria* (Conrad, 1863) (Figure 5.4H) from the subsequent Choptank Formation.

Genus *Torcula* Gray, 1847
Torcula roperorum new species
(Figure 7.5G,H)

Description: Shell large for genus *Torcula,* heavy, massive, elongated; whorls with prominent central depressed band, dividing each whorl into two low, broad cords separated by wide, shallow sulcus; lower (anteriormost) broad cord bifurcated into two small, thin cords by narrow, deep channel (more obvious on early whorls); bifurcated lower cord and central shallow sulcus ornamented with four to six very fine spiral threads; shoulders and edge of base rounded; aperture proportionally large, oval in shape.

Type Material: Holotype, PRI 9912, length (incomplete) 48 mm, width 18 mm.

Type Locality: Eastern side of Urbanna Creek, Urbanna, Middlesex County, Virginia, 2 miles from the State Road 603 bridge, at base of low cliff.

Stratigraphic Range: Confined to the Cobham Bay Member of the Eastover Formation, Messinian Miocene of Virginia.

Etymology: Named for Dr. and Mrs. Clyde (and Ingrid) Roper of Urbanna, Virginia.

Discussion: The new species, the first of its genus to be reported from the Eastover Formation, most closely resembles the Pliocene *Torcula terstriata* (Rogers and Rogers, 1837) from the Sunken Meadow Member of the Yorktown Formation. Although being similar in overall size and shape, *Torcula roperorum* differs from its descendant species in having a much shallower central sulcus and in having lower and less prominent broad cords on each whorl. The Pliocene *Torcula terstriata* also is a more ornamented shell, with several prominent small cords in the central sulcus (see Campbell, 1993; plate 29, figure 293a), while *T. roperorum* is a much smoother shell, with only fine silky threads running through the sulcus area.

Naticoidea
Naticidae
Polinicinae
Genus *Euspirella* new genus

Diagnosis: Small naticids (average length 12 mm) that resemble miniature versions of large *Euspira* (=*Lunatia*) species; shells globose, inflated with rounded but distinctly angled shoulders; spire height variable, with some species having high, scalariform spires and others having low spires; small but deep umbilicus present, with some species having small, thick callus along apertural side of umbilicus; aperture proportionally large, oval shaped.

Type Species*:* *Euspirella chancellorensis* Petuch and Drolshagen, new species (described next), Chancellor Point Sandstone, St. Mary's Formation, early Tortonian Miocene of Maryland.

Other Species in *Euspirella:* *Euspirella tuomeyi* (Whitfield, 1894), Plum Point Member, Calvert Formation of Maryland and Kirkwood Formation of New Jersey and Delaware, Langhian Miocene (Figure 4.5C); *Euspirella hemicrypta* (Gabb, 1860), Choptank Formation of Maryland, Serravallian Miocene (Figure 5.3H); *Euspirella santamaria* Petuch and Drolshagen, new species, Windmill Point Member, St. Mary's Formation, early Tortonian Miocene of Maryland (Figure 6.6G); *Euspirella permeliae* Petuch and Drolshagen, new species, Cobham Bay Member, Eastover Formation of Virginia, Messinian Miocene (Figure 7.3G,H).

Etymology: The genus is named for its resemblance to a small *Euspira* species.

Discussion: This distinctive group of tiny naticids has been referred to by all previous authors as either "*Lunatia,*" "*Euspira,*" or "*Natica*" species. Besides the great difference in size between *Euspirella* and *Euspira* species, the new genus differs from the other large Miocene naticids in having a proportionally much smaller and narrower umbilicus and in having distinctly angled shoulder whorls. *Euspirella* first appears in the late Burdigalian–Langhian Kirkwood and Calvert Formations and makes its last appearance in the Messinian Cobham Bay Member of the Eastover Formation. Throughout its geochronologic range, *Euspirella* was confined to the Transmarian Province and the Chesapeake Miocene.

Euspirella chancellorensis new species
(Figure 6.8C,D)

Description: General description as for diagnosis of genus; shell very small, averaging 9 mm, with angled shoulders and low, flattened spire; sides of body whorl flattened, particularly along edge of aperture, giving shell compressed appearance; umbilicus very narrow and deep, with small callus along apertural edge; aperture proportionally large, oval shaped.

Type Material: Holotype, PRI 9913, length 9 mm, width 9 mm; three specimens in the research collection of the senior author.

Type Locality: In sandstone ledge at high tide line (Chancellor Point Sandstone), Chancellor Point, St. Mary's City, St. Mary's County, on the St. Mary's River, Maryland.

Stratigraphic Range: Restricted to the Chancellor Point Sandstone, St. Mary's Formation, early Tortonian Miocene of Maryland.

Etymology: Named for Chancellor Point, the type locality.

Discussion: *Euspirella chancellorensis,* the type of the genus, is closest to *E. santamaria* Petuch and Drolshagen, new species from the Windmill Point Member of the St. Mary's Formation (Figure 6.6G). The new species differs from the older *E. santamaria* in being a much smaller, more cylindrical shell with a much lower, flattened spire, and in having a smaller and narrower umbilicus.

Euspirella permeliae new species
(Figure 7.3G,H)

Description: General description as for diagnosis of genus; shell globose, inflated, broad across shoulder, with stepped spire; shoulder rounded but distinctly angled; umbilicus proportionally large.

Type Material: Holotype, PRI 9914, length 15 mm, width 14 mm; three specimens in the research collection of the senior author.

Type Locality: In low cliff along the south shore of the Rappahannock River, 5 miles south of Bowler's Wharf, Middlesex County, Virginia.

Stratigraphic Range: At present, known only from the Cobham Bay Member of the Eastover Formation, Messinian Miocene of Virginia.

Etymology: Named for Beth Permelia of Urbanna, Virginia.

Discussion: The new species, the last-living member of its genus, is most similar to *Euspirella santamaria* Petuch and Drolshagen, new species (Figure 6.6G) from the Windmill Point Member of the St. Mary's Formation. *Euspira permeliae* differs from its ancestor in being a smaller, more inflated shell with a proportionally lower spire. The holotype is missing a layer of shell material from around the sutural area above the shoulder, giving it an exaggerated spire height. When this shell loss is factored into the shell proportions, it is obvious that *E. permeliae* has a much lower spire than does *E. santamaria*. In this aspect, the new species shows a close affinity to the low-spired *E. chancellorensis* from the Chancellor Point Sandstone.

Euspirella santamaria new species
(Figure 6.6G)

Description: General description as for diagnosis of genus; shell large for genus, averaging 18 mm, inflated, globose, with distinctly angled shoulder; subsutural area flattened, slightly sloping; spire whorls elevated, protracted, scalariform; umbilicus proportionally small, deep, with small callus along apertural side.

Type Material: Holotype, PRI 9915, length 18 mm, width 15 mm; four specimens in the research collection of the senior author.

Type Locality: In lower shell bed along beach at Chancellor Point, St. Mary's City, St. Mary's County, on the St. Mary's River, Maryland.

Stratigraphic Range: Restricted to the Windmill Point Member (Shattuck Zone 24) of the St. Mary's Formation, early Tortonian Miocene of Maryland.

Etymology: Named for the St. Mary's River.

Discussion: The senior author (Petuch, 2004: plate 36, figure J) misidentified this common St. Mary's species as "*Euspira tuomeyi* (Whitfield, 1894)," a species now known to be restricted to the older Langhian Kirkwood and Calvert Formations (Figure 4.5C, Chapter 4). As discussed in the previous description, *Euspirella santamaria* is closest to *E. permeliae* of the Cobham Bay Member, Eastover Formation of Virginia, but differs in being a narrower, more cylindrical shell with a higher, more protracted spire.

Genus *Neverita* Risso, 1826
Neverita taylorae new species
(Figure 7.4D,E)

Description: Shell small for genus, flattened, broadly discoidal in shape; shoulder rounded, without distinct angle; spire low, compressed, with only earliest whorls projecting; aperture proportionally large, forming ½ of shell base; umbilicus completely filled with large, rounded callus; umbilical callus proportionally large, roughly ¼ of shell base.

Type Material: Holotype, PRI 9916, length 25 mm, width 33 mm; one specimen in the research collection of the senior author.

Type Locality: Eastern side of Urbanna Creek, Urbanna, Middlesex County, Virginia, 2 miles from the State Road 603 bridge, in low cliff.

Stratigraphic Range: Cobham Bay Member of the Eastover Formation, Messinian Miocene of Virginia; may also be present in the Claremont Manor Member of the Eastover Formation.

Etymology: Named for Ms. Beatrice Taylor, Mayor of Urbanna, Virginia.

Discussion: The small, discoidal *Neverita taylorae* is the direct descendant of, and is most similar to, *N. discula* Petuch, 1988 from the Windmill Point Member of the St. Mary's Formation (Figure 6.6C, Chapter 6). The new species differs from its St. Mary's ancestor in being a more inflated and less flattened shell with a more compressed, less projecting spire. *Neverita taylorae* also occurs with *N. vautrotorum* new species (described next), but differs from its sympatric congener in being consistently a much smaller, much more flattened shell with a low, flattened spire.

Neverita vautrotorum new species
(Figure 7.4A,B)

Description: Shell of average size for genus, inflated, globose, with pronounced high conical shape; spire elevated, protracted, with steeply sloping spire whorls; shoulder greatly sloping, running directly into spire whorls; periphery of body whorl rounded, angled along edge of base; aperture proportionally large, flaring, forming ½ of shell base; umbilicus completely filled with large, rounded callus.

Type Material: Holotype, PRI 9917, length 39 mm, width 36 mm; one specimen in the research collection of the senior author.

Type Locality: Eastern side of Urbanna Creek, Urbanna, Middlesex County, Virginia, 2 miles from State Road 603 bridge, in low cliff.

Stratigraphic Range: Cobham Bay Member of the Eastover Formation, Messinian Miocene of Virginia; may also be present in the Claremont Manor member of the Eastover Formation.

Etymology: Named for Mr. and Mrs. James (and Bonnie) Vautrot of Urbanna, Virginia.

Discussion: The large, conical *Neverita vautrotorum* is the direct descendant of, and is most similar to, *N. asheri* Petuch, 1988 from the Windmill Point Member of the St. Mary's Formation (Figure 6.6F, Chapter 6). Both *N. asheri* and *N. vautrotorum* share the same high-spired conical shape, but the new species differs in being more inflated and globose, in having sloping, not stepped, spire whorls, and in having a proportionally higher, more protracted spire. In this last character, *Neverita vautrotorum* has the highest spire of any of the known Chesapeake Miocene *Neverita* species, making it easy to identify in the field.

Muricoidea
Muricidae
Ocenebrinae
Genus *Ecphora* Conrad, 1843
Ecphora amyae new species
(Figure 5.4J,K)

Description: Shell of average size for genus, cylindrical in shape, with high, protracted, scalariform spire, and sculptured with four thick, wide ribs; ribs flattened on edges, slightly "T"-shaped in cross section, ornamented with single large, deeply incised spiral thread, producing thin, shallow groove; shoulder rib extending upward (posteriorward), producing canaliculate spire whorls; subsutural areas of shoulder flat or slightly sloping; areas between ribs ornamented with numerous coarse spiral threads and strong longitudinal growth increments, giving shell rough, scaly appearance; umbilicus proportionally narrow; aperture oval in shape; siphonal canal ornamented with 15 to 17 large spiral cords.

Type Material: Holotype, PRI 9918, length 37 mm, width 33 mm; one specimen in the research collection of the senior author.

Type Locality: Upper bed at Drumcliff, St. Mary's County, on the Patuxent River, Maryland.

Stratigraphic Range: Restricted to the Boston Cliffs Member (Shattuck Zone 19) of the Choptank Formation, Serravallian Miocene of Maryland.

Etymology: Named for Ms. Amy Plummer of Plum Point, Calvert County, Maryland.

Discussion: *Ecphora amyae* is a rare species, known from only a few incomplete specimens. By having slightly "T"-shaped ribs and by having an upward-extending shoulder rib, the new species can be seen to be the descendant of the Shattuck Zone 18 *Ecphora meganae* Ward and Gilinsky, 1988 and represents the last-living member of the *Ecphora meganae* Species Complex. The unrelated sympatric *Ecphora williamsi* Ward and Gilinsky, 1988, a member of the wide-ribbed species group, was originally incorrectly described as a subspecies of *E. meganae*. The discovery of *E. amyae* shows that *E. williamsi* is not the descendant of *E. meganae,* as proposed by Ward and Gilinsky, but is actually a descendant of wide-ribbed forms such as *E. chesapeakensis* Petuch, 1992 and *E. calvertensis* Petuch, 1988.

Genus *Trisecphora* Petuch, 1988
Trisecphora bartoni new species
(Figure 4.6G,H)

Description: Shell of average size for genus, distinctly vase-shaped, with wide shoulder and midbody area tapering gradually to elongated siphonal canal; spire whorls scalariform, with flat subsutural areas; ribs three in number, thin, bladelike, with shoulder rib projecting slightly upward, producing canaliculate subsutural area; fourth small, undeveloped rib present around base of siphonal canal on some specimens; aperture proportionally large, oval in shape; umbilicus narrow and deep; siphonal canal ornamented with numerous faint spiral threads.

Type Material: Holotype, PRI 9920, length (incomplete) 34 mm, width 27 mm; 2 specimens in collection of senior author.

Type Locality: In Shattuck Zone 16, 1/2 mile north of Governor Run, Calvert Cliffs, Calvert County, Maryland.

Stratigraphic Range: Restricted to Shattuck Zone 16 of the Calvert Beach Member, Calvert Formation, latest Langhian Miocene of Maryland.

Etymology: Named in memory of our friend and amateur naturalist, Matthew Barton, late of Boca Raton, Florida.

Discussion: With its distinctive tapering, vase-shaped shell, *Trisecphora bartoni* cannot be confused with any other three-ribbed ecphora species. Small specimens of *Trisecphora smithae* from the St. Leonard Member of the Choptank Formation could be confused with the new species, but are easily separable by having more inflated and globose shells and by having better-developed and larger bladelike ribs.

Genus *Urosalpinx* Stimpson, 1865
Urosalpinx subrusticus bartoni new subspecies
(Figure 6.2E)

Description: Shell of average size for genus, elongated, with high, scalariform spire and long, slender siphonal canal; whorls with 11 to 13 large rounded varices; body whorl heavily sculptured with 8 to 10 large, primary spiral cords, with one or two smaller secondary cords being present between each pair of primary cords; primary cords around edge of shoulder and around base of siphonal canal often largest and most prominent; siphonal canal ornamented with eight large primary spiral cords and 10 smaller secondary cords; all spiral cords intersected by longitudinal growth lines, producing rows of small beads; subsutural area indented and flattened, producing distinctly stepped spire whorls; aperture oval shaped.

Type Material: Holotype, PRI 9921, length 28 mm, width 14 mm; three specimens in the research collection of the author.

Type Locality: Shattuck Zone 22 at Little Cove Point, Calvert County, Maryland.

Stratigraphic Range: Confined to the upper Little Cove Point Member (Shattuck Zone 22) of the St. Mary's Formation, latest Serravallian Miocene of Maryland.

Etymology: Named in memory of our friend and amateur naturalist, Matthew Barton, late of Boca Raton, Florida.

Discussion: This new muricid is described as a Little Cove Point subspecies of the younger *Urosalpinx subrusticus* (d'Orbigny, 1852) (Figure 6.5K) from the Windmill Point Member. *Urosalpinx subrusticus bartoni* differs from the nominate subspecies in being a more elongated and higher-spired shell with distinctly scalariform spire whorls, and in having a much more heavily sculptured shell, ornamented with coarse, variable-sized beaded spiral cords. In the nominate subspecies, the spiral cords are equal in size and are much smoother, giving the shell a silky appearance. At the type locality at Little Cove Point, *Urosalpinx subrusticus bartoni* is commonly encountered, and it must have been a major drilling predator on small bivalves and turritellid gastropods.

Trophoninae
Genus *Patuxentrophon* Petuch, 1992
Patuxentrophon plummerae new species
(Figure 4.2H)

Description: Shell of average size for genus, fusiform, with high, scalariform spire and long, narrow siphonal canal; shoulder sharply angled, ornamented with single large, rounded spiral cord; body whorl heavily sculptured with six large primary cords, with single smaller secondary cord present between each pair of primaries; siphonal canal ornamented with six large primary spiral cords and eight smaller secondary cords; flattened subsutural areas of spire whorls ornamented with three small, fine spiral cords; aperture oval in shape.

Type Material: Holotype, PRI 9919, length 12 mm, width 7 mm.

Type Locality: In Shattuck Zone 10, at base of cliff at Plum Point, Calvert County, Maryland.

Stratigraphic Range: Confined to the Plum Point Member (Shattuck Zone 10) of the Calvert Formation, Langhian Miocene of Maryland.

Etymology: Named for Ms. Amy Plummer of Plum Point, Calvert County, Maryland.

Discussion: The new ancestral Calvert species differs from the descendant Choptank Formation *Patuxentrophon patuxentia* (Martin, 1904) (Figure 5.4G) in being a smaller, more inflated shell with a lower, less protracted spire and in being a much more heavily sculptured shell, with finer and more numerous spiral cords. The shoulder cord of *P. plummerae* is thin, only slightly larger than the other body whorl cords. The shoulder cord of *P. patuxentia,* on the other hand, is much thicker than the other body whorl cords and is much more prominent. The Choptank *Patuxentrophon patuxentia* has four to six large cords on the body whorl, while *P. plummerae* has six main cords and at least six more secondary cords intercalated in between. *Patuxentrophon plummerae* is a rare species, being found only at the type locality at Plum Point.

Buccinoidea
Busyconidae
Busycotypinae
Genus *Busycotypus* Wenz, 1943
Busycotypus martini new species
(Figure 6.2L)

Description: Shell of average size for genus, thin, highly inflated, with protracted, scalariform spire; shoulder sharply angled, with large, thick carina; carina undulating, ornamented with 16 to 17 large, rounded knobs per whorl; subsutural area slightly sloping, ornamented with eight thin, evenly spaced low spiral cords; sutural sulcus proportionally large, deep, bounded by small, thin carina; body whorl ornamented with 20 to 24 low, thin spiral cords, which become nearly obsolete on body whorl of adults; siphonal canal narrow, elongated, ornamented with numerous fine spiral cords and threads.

Type Material: Holotype, PRI 9922, length 178 mm, width 82 mm; two specimens in the research collection of the senior author.

Type Locality: Shattuck Zone 22 at Little Cove Point, Calvert County, Maryland.

Stratigraphic Range: Confined to the upper Little Cove Point Member (Shattuck Zone 22) of the St. Mary's Formation, late Serravallian of Maryland.

Etymology: Named for Dr. George C. Martin, who put together the first compendium of Chesapeake Miocene gastropods in 1904.

Discussion: *Busycotypus martini* is most similar to, and is the direct ancestor of, *B. rugosum* (Conrad, 1843) (Figure 6.3C) from Shattuck Zone 24. The new species differs from its descendant in being a consistently smaller, more inflated shell with a higher, more protracted and scalariform spire and in having distinctly more sloping subsutural areas. *Busycotypus martini* also differs from *B. rugosum* in being a more ornately sculptured shell with 16 to 17 knobs per whorl on the shoulder carina and in having finer and more numerous spiral cords on the body whorl and siphonal canal. *Busycotypus rugosum* has only 12 to 13 shoulder knobs, and these are larger and more rounded than those on *B. martini.* The prominent sutural sulcus of *B. martini* is also larger, wider, and deeper than that of *B. rugosum.*

Nassariidae

Genus *Ilyanassa* Stimpson, 1865

Ilyanassa chancellorensis new species

(Figure 6.8I,J)

Description: Shell small for genus, fusiform, with high, scalariform spire; suture indented, producing distinctly stepped spire whorls; shoulder rounded, angled only at edge of indented suture; body whorl and spire ornamented with 12 large rounded varices per whorl; varices intersected by nine large spiral cords; last pair of cords near siphonal canal doubled, overhanging base of siphonal canal; top of each varix, along indented suture, with large, rounded knob, producing coronated shoulder whorls; siphonal canal short, separated from body whorl by smooth, deeply indented band; aperture proportionally small, oval in shape, with six large teeth along inner edge of lip; columellar fasciole wide, bandlike, with two small cords; outer lip developed into single large varix; protoconch proportionally large, rounded, globose, composed of two whorls.

Type Material: Holotype, PRI 9923, length 10 mm, width 6 mm.

Type Locality: In sandstone ledge at high tide line (Chancellor Point Sandstone), Chancellor Point, St. Mary's City, St. Mary's County, on the St. Mary's River, Maryland.

Stratigraphic Range: Known only from the Chancellor Point Sandstone, St. Mary's Formation, early Tortonian Miocene of Maryland.

Etymology: Named for Chancellor Point, St. Mary's County, Maryland, the type locality.

Discussion: This new nassarid from the Chancellor Point Sandstone is similar only to *Ilyanassa trivittatoides* (Whitfield, 1894) from the Plum Point Member (Shattuck Zone 10) of the Calvert Formation (Figure 4.5M) and from the Kirkwood Formation of southern New Jersey and Delaware. The early Tortonian *Ilyanassa chancellorensis* differs from its Langhian ancestor in being a broader, more inflated shell with a more indented suture and in having finer and more numerous spiral cords. An undescribed species from the Choptank Formation, illustrated by Martin (1904; plate 49, figure 3), also appears to be closely related.

Ilyanassa lindae new species

(Figure 7.3E,F)

Description: Shell elongated, fusiform, with very high, protracted spire; suture impressed, producing scalariform spire whorls; body whorl and spire whorls ornamented with 25 to 27 thin longitudinal ribs; ribs crossed by 10 impressed spiral threads, producing beaded appearance; cord below suture largest and most deeply impressed, separating subsutural areas from body and spire whorls and producing coronated subsutural areas; siphonal canal very short, grading directly into body whorl; columellar fasciole with single large plication; outer lip thin, without varical thickening, smooth along interior; protoconch proportionally large, rounded, composed of 1½ whorls.

Type Material: Holotype, PRI 9924, length 14 mm, width 6 mm; three additional specimens in the research collection of the senior author.

Type Locality: In low cliff along the south shore of the Rappahannock River, 5 miles south of Bowler's Wharf, Middlesex County, Virginia.

Stratigraphic Range: Restricted to the Cobham Bay Member of the Eastover Formation, Messinian Miocene of Virginia.

Etymology: Named for Linda J. Petuch, wife of the senior author.

Discussion: *Ilyanassa lindae* is most similar to, and is the descendant of, *I. peralta* (Conrad, 1830) from the Windmill Point Member (Shattuck Zone 24) of the St. Mary's Formation (Figure 6.6M). The new Messinian species differs from its early Tortonian ancestor in being a smaller, more slender and less inflated shell, with a noticeably finer sculpture pattern on the body and spire whorls. In *Ilyanassa lindae,* there are 25 to 27 ribs per whorl while *I. peralta* has only 18 to 20 ribs per whorl.

Volutoidea
Volutidae
Scaphellinae
Genus *Volutifusus* Conrad, 1863
Volutifusus rappahannockensis new species
(Figure 7.1F,G)

Description: Shell of average size for genus, short and stocky, heavy and thickened, with high, distinctly scalariform, stepped spire; shoulder angled, with distinct sloping subsutural area; spire whorls ornamented with 10 to 15 large, low, equally spaced elongated knobs; shoulder area of body whorl with 10 very low, faint undulating plicae; entire shell covered with thick layer of enamel (chipped away on part of dorsum of holotype); aperture elongated, flaring; columella with two large, prominent plications; siphonal canal short, laterally recurved; calcarella projecting, smooth, rounded on end.

Type Material: Holotype, PRI 9925, length 87 mm, width 45 mm.

Type Locality: Exposed in shallow water at extreme low tide, along the south shore of the Rappahannock River, 5 miles south of Bowler's Wharf, Middlesex County, Virginia.

Stratigraphic Range: Confined to the Claremont Manor Member of the Eastover Formation, late Tortonian of Virginia.

Etymology: Named for the Rappahannock River of Virginia.

Discussion: Of the six described *Volutifusus* species from the Chesapeake Miocene, *V. rappahannockensis* is most similar to, and is the direct descendant of, the common *V. mutabilis* (Conrad, 1834) from the Windmill Point Member (Shattuck Zone 24) of the St. Mary's Formation (Figure 6.4L). The new late Tortonian species differs from its early Tortonian ancestor in being a stockier, squatter, more inflated shell with a more sharply angled shoulder and much more stepped spire whorls. The spire whorls of *V. rappahannockensis* are also more ornamented than those of *V. mutabilis,* being sculptured with large, prominent, undulating knobs. In having a stocky, wide-shouldered shell, the new species also shows some affinity to *V. asheri* Petuch, 1988, also from the Windmill Point Member (Figure 6.4H). *Volutifusus rappahannockensis* differs from *V. asheri* in having a proportionally higher and much more stepped spire and in having prominent spire knobs. The new species is the only known Chesapeake Miocene *Volutifusus* that covers the entire shell with a thick enamel layer.

Cancellarioidea
Cancellariidae
Cancellariinae
Genus *Marianarona* Petuch, 1988
Marianarona gernanti new species
(Figure 5.3I)

Description: Shell of average size for genus, rotund, inflated, very thick and heavy, with proportionally low, only slightly protracted spire; shoulders rounded; body and spire whorls ornamented with 12 thick, low longitudinal ribs; ribs crossed by 10 large, narrow spiral cords; intersection of rib and cord producing large, elongated bead, giving shell rough appearance; aperture proportionally large, flaring, with six large riblike cords in interior; inner edge of lip with six large teeth, corresponding to interior cords; columella with three large teeth and one columellar denticle; small parietal shield present, overhanging small, shallow umbilicus; siphonal canal extremely short, fusing directly into body whorl.

Type Material: Holotype, PRI 9926, length 12 mm, width 9 mm.

Type Locality: In lower bed (Shattuck Zone 18) at Drumcliff, St. Mary's County, on the Patuxent River, Maryland.

Stratigraphic Range: Confined to the Drumcliff Member (Shattuck Zone 18) of the Choptank Formation, early Serravallian Miocene of Maryland.

Etymology: Named for Dr. Robert Gernant, late of the Department of Geology, University of Wisconsin–Milwaukee, who conducted extensive research in the Choptank Formation and who was a mentor of, and inspiration to, the senior author when he was a student at UWM.

Discussion: The new species is most similar to *Marianarona asheri* Petuch, 1988 from the Windmill Point Member (Shattuck Zone 24) of the St. Mary's Formation (Figure 6.7J), but differs in being a much heavier and more globose shell, and in having a much lower, non-scalariform spire. In having the sloping, rounded shoulders and low spire, *M. gernanti* is also similar to *M. reticulatoides* (Martin, 1904) from the Plum Point Member (Shattuck Zone 10) of the Calvert Formation, but differs in having a lower spire and much coarser beaded sculpture.

Marianarona lindae new species
(Figure 6.1I)

Description: Shell of average size for genus, thin, inflated, globose, with sharply angled shoulder and protracted, stepped spire; body whorl and spire whorls ornamented with 15 large, sharp, longitudinal ribs; ribs crossed by 16 thin spiral cords, producing small bead at each intersection; subsutural area flattened, forming stepped, scalariform spire; spiral cord on periphery of shoulder angle largest, forming small shoulder carina; aperture wide, flaring, with seven faint cords in interior; columellar area with three large teeth and one parietal denticle; small parietal shield present, extending over narrow, deep umbilicus.

Type Material: Holotype, PRI 9927, length 17 mm, width 11 mm.

Type Locality: In lower bed (Shattuck Zone 21), ½ mile north of Little Cove Point, Calvert County, Maryland.

Stratigraphic Range: Confined to the lower Little Cove Point Member (Shattuck Zone 21) of the St. Mary's Formation, late Serravallian Miocene of Maryland.

Etymology: Named for Linda J. Petuch, wife of the senior author.

Discussion: The new species is most similar to, and is the probable ancestor of, *Marianarona alternata* (Conrad, 1834) of The Windmill Point Member (Shattuck Zone 24) of the St. Mary's Formation (Figure 6.7F). *Marianarona lindae* differs from its descendant in being a more inflated shell with a more sloping, rounded subsutural area and in having a proportionally much lower, less exserted spire. The early Tortonian *M. alternata* is also a much more coarsely sculptured shell than is the late Serravallian *M. lindae,* having only 10 large, thick spiral cords on the body whorl as opposed to the 16 fine, threadlike cords of the new species.

Bivalvia
Pteriomorpha
Arcoida
Arcoidea
Arcidae
Anadarinae
Genus *Dallarca* Ward, 1992
Dallarca chancellorensis new species
(Figure 6.8K,L)

Description: Shell of average size for genus, slightly inflated, roughly rectangular in shape, with proportionally large, nearly equal-sized lateral auricular extensions on either side of beaks; posterior end projecting only slightly; hinge plate straight, with 55 small chevron-shaped teeth; break in dentition direction anterior of shell midline; beaks and umbonal area proportionally low for genus, extending only slightly past hinge line; shell exterior sculptured with 30 large, flattened ribs; ribs intersected by extremely numerous fine concentric growth lines, giving ribs faint beaded appearance; inner edge of commisure with 30 large, flattened teeth.

Type Material: Holotype, PRI 9928, length 51 mm, width 41 mm.

Type Locality: In sandstone ledge at high tide line (Chancellor Point Sandstone), at Chancellor Point, St. Mary's City, St. Mary's County, Maryland, on the St. Mary's River.

Stratigraphic Range: Confined to the Chancellor Point Sandstone, St. Mary's Formation, early Tortonian Miocene of Maryland.

Etymology: Named for Chancellor Point, St. Mary's County, Maryland, the type locality.

Discussion: The new Chancellor Point Sandstone species is most similar to *Dallarca idonea* (Conrad, 1832) from the underlying Windmill Point Member (Shattuck Zone 24) (Figure 6.4M), but differs in being a much broader, less inflated shell with a much wider and better developed hinge line with wider and more pronounced auricular extensions. *Dallarca chancellorensis* also differs from *D. idonea* in being a much more rectangular shell, in having much lower, less-projecting beaks and a more compressed umbonal area, and in having more numerous and finer radiating ribs (30 on *D. chancellorensis* and 25 to 27 on *D. idonea*). With its rectangular shape and low beaks, the new species is also similar to *Dallarca rotunda* Ward, 1992 from the Claremont Manor

Member of the Eastover Formation (Figure 7.1E), but differs in being wider and more rectangular in shape and in being more heavily sculptured, with 30 ribs on *D. chancellorensis* and only 22 to 25 on *D. rotunda*.

Dallarca covepointensis new species
(Figure 6.1 M)

Description: Shell small for genus, somewhat flattened, rounded, slightly rectangular in shape; posterior half of shell projecting; hinge line straight, with 42 small, chevron-shaped teeth; break in direction of teeth slightly anterior to shell midline; beaks and umbonal area proportionally low for genus, projecting only slightly beyond hinge line; shell exterior sculpted with 30 large, flattened ribs; ribs intersected by numerous fine concentric growth lines, producing beaded appearance on ribs; inner edge of commisure with 26 wide, flattened teeth.

Type Material: Holotype, PRI 9929, length 48 mm, width 34 mm.

Type Locality: In lower bed (Shattuck Zone 21), ½ mile north of Little Cove Point, Calvert County. Maryland.

Stratigraphic Range: Confined to the Little Cove Point Member (both Zones 21 and 22) of the St. Mary's Formation, late Serravallian Miocene of Maryland.

Etymology: Named for the Little Cove Point Member, to which the new species is restricted.

Discussion: The new Little Cove Point species is most similar to, and is the direct ancestor of, *Dallarca idonea* (Conrad, 1832) from the overlying Windmill Point Member (Shattuck Zone 24) (Figure 6.4M). *Dallarca covepointensis* differs from its descendant in being a much smaller shell with a more rectangular outline, by having distinctly lower, less projecting beaks and a more flattened umbonal area, and by having more numerous and finer ribs (30 on *D. covepointensis* and 25 to 27 on *D. idonea*). Ward (1992; 58, plate 11, figures 1,5) noticed the consistent differences between *D. idonea* and *D. covepointensis* but considered the new species to be only a small subspecies of *D. idonea*.

Pterioida
Pterina
Pterioidea
Isognomonidae
Genus *Isognomon* Lightfoot, 1786
Subgenus *Hippochaeta* Philippi, 1844
Isognomon (Hippochaeta) sargenti new species
(Figure 7.3A,B)

Description: Shell large for subgenus, averaging 240 mm, very elongated and mytiliform, slightly convex, with extremely thick hinge area; beaks strongly recurved anteriorly, arcuate, pointed, with rounded tips; hinge ligament plate proportionally wide with 18 to 19 wide, straight, shallow, evenly spaced byssal grooves; byssal groove interstices broad, concave on surface, extending slightly into body cavity; ventral edge very thickened, distinctly inrolled; outer surface with thin, finely crenulated shell layer that covers main nacreous layer; lower end of shell thin and fragile; commisure smooth.

Type Material: Holotype, PRI 9930, length (pair) 242 mm, width 95 mm; five specimens (hinge areas only), in the research collection of the senior author.

Type Locality: Eastern side of Urbanna Creek, Urbanna, Middlesex County, Virginia, 2 miles from the State Road 603 bridge, at base of low cliff, 2 feet above the tide line.

Stratigraphic Range: Confined to the Claremont Manor and Cobham Bay Members of the Eastover Formation, late Tortonian and Messinian Miocene of Virginia.

Etymology: Named for Dennis Sargent of Leesburg, Florida, who collected the holotype.

Discussion: The new Eastover species is most similar, both in size and shape, to *Isognomon (Hippochaeta) torta* (Say, 1820) (Figure 5.3E) from the Choptank Formation of Maryland (misidentified by Glenn (1904; 383) as "*Melina maxillata* (Deshayes)," a European species). *Isognomon (Hippochaeta) sargenti* differs from its Choptank ancestor in having a more recurved, arcuate, and elongated shell, and in having a proportionally thicker and heavier hinge area. The hinge ligament areas also differ between the two species, with *I. (Hippochaeta) sargenti* having 18 to 19 byssal grooves and *I. (Hippochaeta) torta* having 20 to 24 byssal grooves. The byssal grooves and byssal groove interstices on the new species are also proportionally wider than those of its Choptank ancestor. *Isognomon (Hippochaeta) sargenti* is the last-living member of its subgenus, a victim of the late Messinian extinction event.

Ostreoidea
Ostreidae
Ostreinae
Genus *Mansfieldostrea* new genus

Diagnosis: Oyster shells of average size for family, moderately heavy and thickened, oval in shape; bottom valve very convex, cuplike, heavily ornamented with concentric rows of large fluted fronds and frills, often developed into wide flanges on either side of hinge; upper valve flat or slightly concave, oval in shape, ornamented with simple fine concentric growth lamellae; juveniles attached to substrate but adults unattached, free-living, lying in boatlike fashion on muddy or sandy substrates; hinge simple, with large triangular central ligament scar.

Type Species: "*Ostrea*" *compressirostra* Say, 1824, Sunken Meadow and Rushmere Members, Yorktown Formation of Virginia and northern North Carolina, Duplin and Raysor Formations of North and South Carolina, Jupiter Waterway and Sarasota Members, Murdock Station Formation of southern Florida, and the lower beds of the Jackson Bluff Formation of northwestern Florida, all Zanclean and early Piacenzian Pliocene.

Other Species in *Mansfieldostrea*: *Mansfieldostrea brucei* (Ward, 1992), Claremont Manor Member, Eastover Formation of Virginia and the Cocoplum Member, Bayshore Formation of southern Florida, late Tortonian Miocene; *Mansfieldostrea geraldjohnsoni* (Ward, 1992), Cobham Bay Member, Eastover Formation of Virginia and the Port Charlotte Member, Bayshore Formation of southern Florida, Messinian Miocene; *Mansfieldostrea disparilis* (Conrad, 1840), Mogarts Beach and Moore House Member, Yorktown Formation of

Virginia and northern North Carolina, upper Duplin Formation of North and South Carolina, Pinecrest Member, Tamiami Formation of southern Florida; *Mansfieldostrea raveneliana* (Tuomey and Holmes, 1855), Chowan River Formation of Virginia and North Carolina, Fruitville Member, Tamiami Formation of southern Florida, latest Piacenzian Pliocene.

Etymology: Named as a combination of "*Ostrea*" ("oyster") and Dr. Wendell Mansfield, who through his many publications during the 1920s to 1940s, contributed so much to our knowledge of the Miocene and Pliocene mollusca of the eastern United States.

Discussion: Until now, this group of large and abundant Miocene and Pliocene American oysters had been considered to belong to the genus *Ostrea* Linnaeus, 1758. Members of *Mansfieldostrea* differ from the predominantly Old World genus *Ostrea* in being far more regularly shaped with laterally symmetrical shells, in having the bottom valve extremely concave while the top valve is consistently smaller and flat, and in being unattached and free-living as adults. The extreme difference between the sculpture patterns of the unornamented upper valve and the frilled and lamellose lower valve also sets *Mansfieldostrea* apart from the attached genus *Ostrea,* with its irregularly shaped shell and sculpture.

Heteroconchia
Veneroida
Crassatelloidea
Astartidae
Astartinae
Genus *Astarte* Sowerby, 1816
Astarte lindae new species
(Figure 7.1L)

Description: Shell of average size for genus, subtrigonal in shape, thickened, slightly inflated; posterior slope slightly convex, gently rounded; anterior slope slightly concave; beak broad, recurved anteriorly, ornamented with at least 10 thin, fine, concentric undulations (beak of holotype eroded; at least another eight undulations would be present on a perfect specimen); posterior ¾ of external surface smooth, with faint growth lines; anterior ¼ of shell ornamented with thin, evenly spaced concentric ribs, essentially extensions and magnifications of faint growth lines on posterior ¾ of shell exterior; some anterior growth lines as strong as undulations on beaks; anterior and posterior adductor muscle scars of equal size and equally impressed; pallial line curved, without sinus; dentition with two strong cardinal teeth, both equal in size and development; lunule elongate; inner margin of commisure crenulate.

Type Material: Holotype, PRI 9931, length 21 mm, width 24 mm.

Type Locality: Eastern shore of Urbanna Creek, Urbanna, Middlesex County, Virginia, 3 miles from State Road 603 bridge, in lower bed of low cliff, at low tide line.

Stratigraphic Range: Confined to the Claremont Manor Member, Eastover Formation, late Tortonian Miocene of Virginia.

Etymology: Named for Linda J. Petuch, wife of the senior author.

Discussion: The new Claremont Manor species is most similar to, and is the ancestor of, the common Cobham Bay Member *Astarte rappahannockensis* Gardner, 1944 (Figure 7.2J). The late Tortonian *Astarte lindae* differs from its Messinian descendant in being a broader, more trigonal shell with a proportionally much wider beak and umbonal area. The new species is also a more ornate shell, with finer and more numerous undulations on the beaks (approximately 18 on *A. lindae* and only 10 on *A. rappahannockensis*) and in having a split sculpture pattern with strong, distinct riblets on the anterior ¼ of the shell exterior. In contrast, *A. rappahannockensis* is completely smooth over the entire shell disc below the beak undulations.

Carditoidea
Carditidae
Carditamerinae
Genus *Cyclocardia* Conrad, 1867
Cyclocardia calvertensis new species
(Figure 4.1D)

Description: Shell small for genus, inflated, ovately subtrigonal, thick, with projecting beak; anterior slope slightly concave; posterior slope rounded, convex; shell exterior ornamented with 20 large, wide concentric ribs, becoming thicker and coarser toward anterior end; ribs crossed by evenly spaced growth lines, producing large, elongated bead at each intersection; umbones strongly recurved, extending onto lunule; cardinal tooth of right valve (holotype) broadly "V"-shaped; inner edge of commisure margin with 18 large, flattened teeth.

Type Material: Holotype, PRI 9932, length 12 mm, width 12 mm; one specimen attached to the holotype of *Xenophora lindae* new species.

Type Locality: In Shattuck Zone 10 of the Plum Point Member at Plum Point, Calvert County, Maryland.

Stratigraphic Range: Confined to the upper Plum Point Member (Shattuck Zone 10) of the Calvert Formation, Langhian Miocene of Maryland.

Etymology: Named for the Calvert Formation.

Discussion: The new species was referred to as "*Venericardia granulata* Say" by Glenn (1904, plate 91, figures 9,10), but that taxon is now known to represent a Pliocene species from the Yorktown Formation (see Campbell, 1993; 34). The Langhian *Cyclocardia calvertensis* differs from the Piacenzian Pliocene *C. granulata* in being a much smaller, more rounded, and less elongated shell with fewer and coarser ribs (20 on *C. calvertensis* and 22 on *C. granulata* (an average between 19 and 25)). The ribs on the new species are also more coarsely sculptured than those on *C. granulata,* having proportionally larger and fewer beads. Another, larger undescribed species from the St. Mary's Formation was also illustrated by Glenn (1904: 91, figures 7,8).

Cyclocardia vautrotorum new species
(Figure 7.4C,F)

Description: Shell large for genus, elongately ovate in shape, heavy and thickened; beak projecting, strongly recurved anteriorly; anterior slope slightly concave; posterior slope very rounded, convex; shell exterior ornamented with 26 low, broad, flattened radial ribs; ribs curved anteriorly, stronger on umbones and first 1/3 of shell, becoming gradually more flattened and nearly obsolete on last 2/3 of shell; cardinal teeth of holotype (right valve) composed of single large, long tooth and a corresponding small tooth arranged in lopsided "V"-shaped pattern; anterior adductor muscle scar larger than posterior, elongately oval in shape; pallial line simple, curved, deeply impressed; inner edge of commisure with 18 large, flattened teeth.

Type Material: Holotype, PRI 9933, length 25 mm, width 24 mm; four specimens in research collection of the senior author.

Type Locality: Eastern shore of Urbanna Creek, Urbanna, Middlesex County, Virginia, 2 miles from State Road 603 bridge, in base of low cliff.

Stratigraphic Range: Known only from the Cobham Bay Member of the Eastover Formation, Messinian Miocene of Virginia.

Etymology: Named for Mr. and Mrs. James (and Bonnie) Vautrot of Urbanna, Virginia.

Discussion: With its low, flat, almost obsolete ribs and nearly smooth exterior shell surface, the new species is similar only to *Cyclocardia castrana* (Glenn, 1904) from the Plum Point Member of the Calvert Formation. The Messinian *Cyclocardia vautrotorum* differs from the Langhian *C. castrana* in being a much larger and more elongated shell with more flattened ribs and a smoother appearance. The juvenile of *C. vautrotorum* (Figure 7.4F) is much more similar to the ancestral *C. castrana,* having stronger, slightly beaded ribs. This similarity is only ontogenetic, as the ribs of the new species flatten out and almost disappear on older specimens. The flat-ribbed *Cyclocardium vautrotorum* is also similar to the Chattian Oligocene–Aquitanian Miocene *C. trentensis* Ward, 1992 from the Belgrade Formation of northern North Carolina and, together with *C. castrana,* they form a distinct smooth-surfaced species group.

Veneridae
Chioninae
Genus *Mercenaria* Schumacher, 1817
Mercenaria druidi gardnerae new subspecies
(Figure 7.1K)

Description: Shell small for genus, thick, heavy, broad, laterally elongated, roughly rectangular in shape; beak and umbones highly recurved anteriorly; shell exterior sculptured with very numerous growth lamellae that fuse together to form broad, flat surfaces; in weathered and decorticated specimens, fused growth lamellae absent, leaving exposed under layer sculpted with extremely fine radial riblets; large, widely spaced growth increments are also preserved on decorticated specimens (as in holotype); escutcheon proportionally small and poorly developed; anterior and posterior adductor muscles roughly same size, oval in shape; pallial line deeply incised, with large, "V"-shaped pallial sinus at posterior end; dentition of right valve with two large bifurcated cardinal teeth and one small, thin posterior cardinal; inner edge of commisure margin ornamented with numerous fine cenulations.

Type Material: Holotype, PRI 9934, length (pair) 68 mm, width 81 mm; four specimens in the research collection of the author.

Type Locality: Eastern shore of Urbanna Creek, Urbanna, Middlesex County, Virginia, 3 miles from State Road 603 bridge, in lower bed of low cliff, at low tide line.

Stratigraphic Range: Confined to the Claremont Manor Member of the Eastover Formation, late Tortonian Miocene of Virginia.

Etymology; Named for Dr. Julia Gardner, who, through her numerous large papers, contributed greatly to our knowledge of the Miocene and Pliocene mollusks of Virginia, the Carolinas, and Florida.

Discussion: The new Claremont Manor venerid is described as a stratigraphic subspecies of the Cobham Bay *Mercenaria druidi* Ward, 1992 (Figure 7.4J). That Messinian species was originally described as "*Venus (Mercenaria) berryi*" by Gardner in 1944, but the taxon was later found to be preoccupied by "*Venus berrii* Wood, 1828." Ward (1992; 107) proposed *Mercenaria druidi* as a replacement name for Gardner's *berryi,* citing the specimens illustrated by Gardner in 1944 (plate 21, figures 1–6). These specimens are definitely of the laterally compressed shell type exhibited by Cobham Bay specimens, especially those from the Bowler's Wharf area. This leaves the older Claremont Manor form without a name. *Mercenaria druidi gardnerae* differs from its nominate subspecies in being a much wider and broader shell with lower, flatter beaks. *Mercenaria druidi druidi* is a much more laterally compressed shell with high projecting beaks. Both subspecies share the same type of shell sculpture, with strong growth lamellae that frequently fuse to form wide, flat areas.

Mercenaria lindae new species
(Figure 6.1K)

Description: Shell small for genus, thick, heavy, laterally compressed, subtrigonal, with high projecting beak; umbones highly recurved anteriorly; escutcheon proportionally small, overhung by umbones; exterior of shell heavily ornamented with very numerous concentric lamellae which frequently fuse into wide, flat areas, giving shell a smooth appearance; lamellae along anterior and posterior edges separate, very frilly and ornate, fusing into wide flat areas as they extend across disc; anterior adductor muscle scar larger than posterior, lunate in shape; pallial line deeply impressed, with "V"-shaped pallial sinus at posterior end; hinge dentition of holotype (left valve) with large anterior cardinal tooth, large bifurcated medial cardinal tooth, and long, bladelike posterior cardinal tooth.

Type Material: Holotype, PRI 9935, length 65 mm, width 74 mm.

Type Locality: In lower bed (Shattuck Zone 21), 1/2 mile north of Little Cove Point, Calvert County, Maryland.

Stratigraphic Range: Found in both Shattuck Zones 21 and 22 of the Little Cove Point Member, St. Mary's Formation, late Serravallian Miocene of Maryland.

Etymology: Named for Linda J. Petuch, wife of the senior author.

Discussion: The new species is most similar to *Mercenaria tetrica* (Conrad, 1838) from the upper Little Cove Point Member (Shattuck Zone 22) and the Windmill Point Member (Shattuck Zone 24) of the St. Mary's Formation (Figure 6.2N). *Mercenaria lindae* differs from *M. tetrica* in being a smaller, more laterally compressed shell with higher, more protracted beaks and in having much smoother shell sculpture, composed of fused concentric growth increments. The fused growth increments, with their shiny, flattened surfaces, are never present on the more roughly sculptured *M. tetrica*. This same type of fused, smooth shell sculpture is also seen in the younger Eastover Formation *Mercenaria druidi gardnerae* new subspecies and *M. druidi* Ward, 1992.

Index

D

E

F

G

H

I

K

L

M

O

P

R

S

T

U

V

X